Mushroom Pest and Disease Control

A Colour Handbook

John T. Fletcher B.Sc, Ph.D
Regional Advisory Pathologist, Agricultural Development and Advisory Service, UK

Richard H. Gaze B.Sc
National Mushroom Specialist, Agricultural Development and Advisory Service and Horticultural Research International, UK

With contributions from

P. F. White, HNC, LIBiol Horticultural Research International, UK
Professor Danny Rinker, University of Guelph, Ontario, Canada
Professor Albert Eicker, University of Pretoria, South Africa
Dr Helen Grogan, TEAGASC, Kinsealy Research Centre, Dublin, Ireland

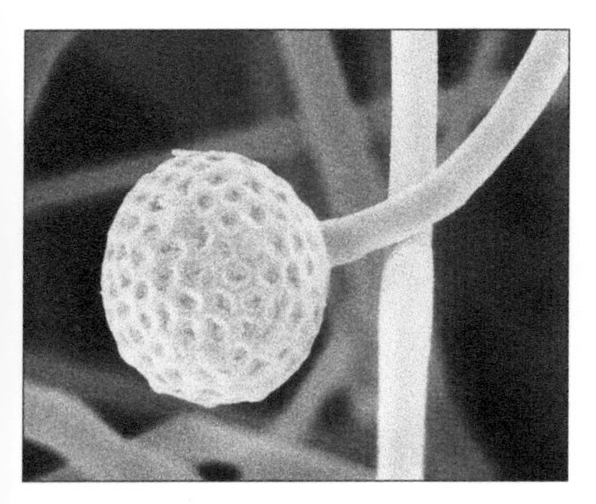

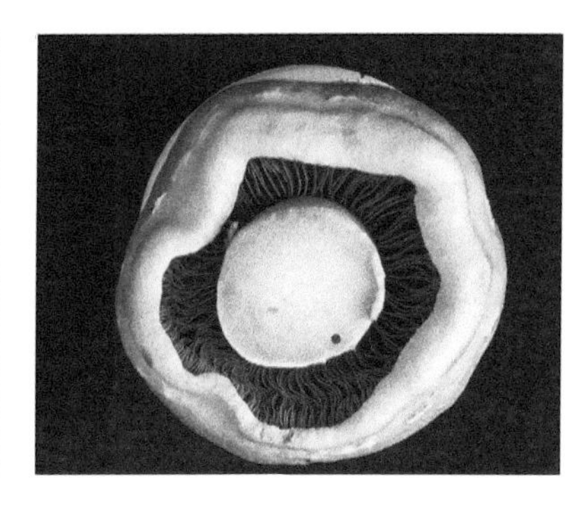

CRC Press
Taylor & Francis Group
Boca Raton London New York

CRC Press is an imprint of the
Taylor & Francis Group, an **informa** business

CRC Press
Taylor & Francis Group
6000 Broken Sound Parkway NW, Suite 300
Boca Raton, FL 33487-2742

Visit the Taylor & Francis Web site at
http://www.taylorandfrancis.com

and the CRC Press Web site at
http://www.crcpress.com

Plant Protection Handbooks Series

Alford: *Pests of Fruit Crops – A Colour Handbook*
Alford: *Pests of Ornamental Trees, Shrubs and Flowers – A Colour Atlas*
Biddle & Cattlin: *Pests and Diseases of Peas and Beans – A Colour Handbook*
Blancard: *Cucurbit Diseases – A Colour Atlas*
Blancard: *Tomato Diseases – A Colour Atlas*
Blancard: *Diseases of Lettuce and Related Salad Crops – A Colour Atlas*
Fletcher & Gaze: *Mushroom Pest and Disease Control – A Colour Handbook*
Helyer *et al*: *Biological Control in Plant Protection – A Colour Handbook*
Koike *et al*: *Vegetable Diseases – A Colour Handbook*
Murray *et al*: *Diseases of Small Grain Cereal Crops – A Colour Handbook*
Wale *et al*: *Pests & Diseases of Potatoes – A Colour Handbook*
Williams: *Weed Seedlings – A Colour Atlas*

Contents

About the Authors

John Fletcher graduated in Horticultural Botany at the University of Reading followed by a PhD in plant pathology at the University of Birmingham. He joined the National Agricultural Advisory Service (subsequently the Agricultural Development and Advisory Service) and worked on many diseases during his 34 years but specialized in those of glasshouse and mushrooms crops. In 1969 he spent a year as a research fellow at the University of Guelph, Ontario, Canada where he researched tomato diseases but also made the initial discovery of the effectiveness of the benzimidazole fungicides for the control of mushroom pathogens. He is a Past President of the British Society of Plant Pathologists, received the British Crop Protection Councils Award for services to British crop protection and the Mushroom Growers Association Sinden Award for his contribution to research and advice on the control of mushroom diseases. In recent years he has been consulted by mushroom growers in the UK and other countries and spent 5 months at La Trobe University, Melbourne, Australia working on various mushroom pathogens and moulds.

Richard Gaze graduated in Horticulture at Wye College of the University of London. Following initial jobs in the protected crops industry in Denmark, he joined the National Agricultural Advisory Service and developed a special interest in the mushroom crop. This lead to him becoming the Agricultural Development and Advisory Service National Mushroom Specialist, a position he held for 20 years and subsequently for a further 6 years when he transferred to Horticulture Research International at Wellesbourne. He produced a regular News Letter on the mushroom industry for ADAS and in 2000 he became the editor of the Mushroom Growers Associations Mushroom Journal, a position he held until its closure in 2006. He was awarded the MGA Sinden Award for his services to the Industry both as an adviser and researcher. He is still involved in mushroom consultation work and advises ADAS as well as being an Associate Fellow of Warwick HRI and member of the Horticultural Development Council Mushroom Panel.

Preface

Mushrooms are a very important crop and are grown commercially in many countries. China grows more types of mushrooms than any other country and is the largest overall producer accounting for some 32%. Worldwide, *Agaricus bisporus*, the Paris mushroom, button mushroom or white mushroom (the subject of this book), is probably the most widely grown. It is eaten both fresh and processed, especially canned. Modern production methods are highly mechanized, requiring detailed knowledge and a high level of management skill for successful and continuous cropping. Inevitably investment and production costs are high and returns have not kept pace with inflation, though the crop is still one of the most valuable in many of the countries where it is grown. In the USA alone in 2003–2004, the *Agaricus* crop was valued at 880 million dollars and was produced by only 125 growers. Twenty-six of these, who produced over 10 million pounds each (c. 4.5 million kg), accounted for 65% of the total production. The situation in Europe is similar where there is also a trend for the numbers of producers to decrease whilst farm size increases.

Fifteen years have passed since the publication of the second edition of the book *Mushrooms: Pest and Disease Control* (published by Intercept, Andover, Hants, England) and this is a major revision of that book. During this time the mushroom industry, worldwide, has undergone many changes. In the USA, fresh mushroom production has increased by 14% over the last 10 years, while processed mushrooms have decreased by a similar amount. In Europe, there has been a significant shift in the importance of the crop. In Poland for instance, production has increased considerably whereas in the UK it has decreased by over 50% in the last 5 years. Cultural developments have continued at a rapid pace, and the production of compost in bulk, up to and including fully colonized compost, is now common practice. The use of bins or bunkers in the early stages of composting has had major effects, both culturally, and on the biology of the process. Finer grades of peat are used in casing, and sugar beet lime, a by-product of sugar production from sugar beet, is very commonly used as the lime source. These changes have resulted in wetter casing, and in turn, this can have a major effect on disease incidence. Wetter casing is considered by some to be one of the main factors in the international increase in cobweb disease. Bulk production of compost and its distribution to growers has accentuated the Trichoderma compost mould problem.

The discovery of Mushroom virus X disease, which shows many of the features of La France disease but with an apparent absence of virus particles, has resulted in a renewed interest in 'virus diseases' of the crop. Research facilities are essential if such new diseases are to be understood and controlled. Unfortunately there has been decreasing support for research and extension in many countries, largely as a result of the withdrawal of government funding. The global industry urgently needs to reassess its research and extension requirements in order to make the best use of decreasing international resources.

Pesticide (fungi and insects) resistance has affected the control of sciarid flies and diseases such as Cobweb and Verticillium. The numbers of pesticides available to mushroom growers has decreased and is likely to decrease further. Biological control of pests and pathogens has not developed at the pace predicted some years ago. There are still no biologically active products available for the control of fungal pathogens, although sciarid control with nematodes is now an established practice.

In addition to all these problems and changes is the growing public concern for the environment. Compost odours, run-off water from compost yards, the use of disinfectants and pesticides, the possible contamination of water courses near to farms, have all added to perceived environmental problems associated with mushroom production. Industries in many countries have addressed many of these potential problems but at a cost. Returns for the product sometimes barely cover costs, and in countries where the crop is mainly marketed through supermarkets there is constant pressure to improve quality, while returns remain constant or even in decline. The high cost of harvesting continues to be a major factor and has resulted in the use of cheap labour as well as the expansion of the industry into countries where labour is less expensive.

Against this background, effective pest and disease control is essential if farms are to remain viable. The aim of this book is the same as that in *Mushrooms: Pest and Disease Control*, namely to provide information on the problems of the crop and the best ways to overcome them to remain in business. In this respect, all the chapters have been revised, some in a major way, and many colour pictures added. Inevitably there is some duplication, for instance in methods of pest and disease control, where the same or similar points are made for various problems. However, in a new Chapter 3 details of processes and procedures with checklists are given, and referred to more briefly under specific problems.

In order to preserve and get the best results from the decreasing number of pesticides available to the industry, it is vital that growers are totally familiar with the label recommendations; not only those that refer to biological efficacy but also the safety instructions. The industry may one day have to manage without pesticides, and the development of pest- or pathogen-resistant strains of spawn still seems to be a long way off. For these reasons alone, it is vital that growers use all means available to them in the management of pest and pathogen populations.

Finally, a word of caution: the international situation regarding the registration of pesticides for use on the crop is ever-changing. There is also no universal agreement on which pesticides should or should not be registered. Each country makes its own decision. With the present speed of change, it is likely that some active ingredients mentioned in this book will have been withdrawn from use in some countries by the time the book is published. In addition, the rate of use of formalin suggested in this book is the maximum permissible rate for the UK; this rate may not be permissible in other countries. It is therefore vitally important that the national list of approved products is consulted before a choice of chemical is made and that it should be used at the rate shown on the label. The product labels must be read very carefully before using a chemical and the stated safety recommendations must be strictly followed at all times. The most toxic pesticides and Formalin require the operator to wear full protective clothing including the correct respirator filter. The product label also gives a harvest interval and this, together with the correct rate, will insure that the treated mushrooms do not have undesirable residues.

It is hoped that this book will be of value to the mushroom industry in every country where the crop is grown. The metric units used in this book are not in common usage in all countries and with this in mind Appendix 3, which contains conversions to other units, has been included.

We express our appreciation to the many people who have contributed in one way or another to the information contained in this book. In particular, we wish to thank Dr Helen Grogan for her very significant research contribution during the past 12 years. We also thank Judy Allan, John Burden, Alan Clift, F.J. Gea, Geoff Izard, Martmari van Greuning, Brian Oxley, Steve Newton, Peter Romaine, Greg Seymour, and Andrew Tinsley for information and photographs and in particular Dr Peter Mills for the use of photographs (78, 101, 122, 130–140, 146–150, 156, and 159) that belonged to the Glasshouse Crops Research Institute and Warwick/HRI. We are also extremely grateful to Professor Fred Last and Pat Fletcher for their very useful comments on, and help with the preparation of the manuscript.

Finally we are very grateful for the contributions which are also acknowledged in the appropriate places in the book, by Professor Danny Rinker for his North American perspective, to Jane Smith for help with the pest section, and to Professor Albert Eicker for his contribution to the moulds chapter.

John T. Fletcher
Richard H. Gaze

CHAPTER 1

Mushroom Growing

- INTRODUCTION
- CULTURE
  Compost: the ingredients; Composting; Compost smells; Compost analysis; Mushroom compost: the selective medium; Spawning, spawn-running and phase III compost; Casing; Cropping; Pests and pathogens
- MUSHROOM GROWING SYSTEMS
  Tray systems; Shelf systems; Bag and block systems; Deep trough system
- BUILDINGS

Introduction

Mushroom culture is a remarkable system of biological manipulation whereby the organisms that are most likely to be harmful are minimized, and those that are beneficial are encouraged. A suitable medium, the compost, is the end product of a complex but controlled biological process involving fungi, bacteria, and actinomycetes. When well prepared, it is a living ecosystem that is suitable for the growth of mushrooms. Mushroom mycelium, once introduced into the compost, affects the system substantially and the development of other microorganisms may be minimized by competition and probably antagonism. However, mushroom compost is not a selective medium in the strict sense, and other fungi introduced at the completion of composting and before mushroom spawn may also grow well, often at the expense of mushroom mycelium.

This book is almost entirely about the white mushroom, *Agaricus bisporus* (also known as *A. brunnescens*), with occasional reference to the closely related species *Agaricus bitorquis*. Many readers will have an intimate knowledge of mushroom growing and the different production systems used. This chapter is included for those without such knowledge. A brief description is given of the production processes, the different systems and the overall environment in which mushrooms grow. An additional aid to understanding is the inclusion of a glossary of terms on pages 178–181.

Culture

Compost: the ingredients

The predominant raw material for mushroom compost, or mushroom substrate as some prefer to call it, is straw. Wheat straw is generally used, although straw from other crops, such as barley, rice and oil seed rape, is also suitable. In the USA, hay is a major ingredient of mushroom compost. Traditionally, mushroom producers obtained wheat straw as horse litter from stables. Baled straw mixed with poultry manure called 'synthetic' compost, is widely used as a substitute for horse litter. More complex mixtures include crushed corncobs, cotton seed meal and fertilizers such as urea or ammonium nitrate which provide additional carbohydrate and nitrogen.

Composting

The process of changing these mixtures into a suitable medium for mushroom production by composting, or fermentation, takes place in distinct phases. Initially the ingredients are mixed and wetted (phase 0), composting begins (phase I), it is pasteurized and composting completed (phase II), and finally it is colonized by mushroom mycelium (phase III).

Prewet or phase 0

The purpose of this phase is to mix and wet the raw material to begin the composting process during which various microorganisms break down the straw. With the increased use of baled straw, a prewetting and blending phase has become more common. During this process the raw material is thoroughly wetted and made into large heaps which are moved frequently (1). This initial wetting and mixing phase occurs over a period of 7 days.

Phase I

After the wetting and mixing phase, the compost is made into long narrow stacks or windrows in which the composting process continues. Traditionally the windrows of phase I compost are made up in the open or under the protection of an open-sided shed (2). The phase I process takes a further 7 days. The centre of a compost stack commonly reaches temperatures (70–75°C), high enough to kill pests and pathogens in the manure or straw. The material is turned several times by mechanical compost turners, often every other day, but some of the outer layers may not reach the middle and consequently do not achieve high temperatures. Within the last decade or so, the use of specially built bunkers with underfloor ventilation, and sometimes with an open or partially open top or under a roof, has become widespread (3). The compost is put into these bunkers after a short prewet. The continuous supply of air from below and the insulation provided by the walls result in the bulk of the compost reaching temperatures of 80°C or more. It is normal to take the compost out of the bunkers at 2- or 3-day intervals and put it back, thus allowing additional mixing. Generally, phase I bunker compost is produced in two or three fewer days (i.e. 11–12 days compared with 14 days) than that produced by the traditional methods.

Prewet/phase I composting methods are still being actively developed to further improve productivity and reduce odour pollution.

Large quantities of water are used in the two early stages of compost production and excess water is generally collected in large containers. The water contains large amounts of organic material which ferments, especially in hot weather. It then becomes anaerobic, producing unpleasant smells. This water, often referred to as 'goody water,' must be well aerated and it can then be recycled and used to make more compost. It can contain large quantities of

1 Straw bale breaker with a chicken manure hopper on a mixing line at the beginning of the prewet operation.

2 Traditional phase II windrows.

3 Phase I bunker with underfloor ventilation holes in the lines in the concrete.

soluble salts which may inhibit composting or mushroom mycelial growth if their concentration becomes too high.

Phase II, peak-heating, pasteurization or sweat-out

The composting process is continued in this phase until it is judged to be suitable for the growth of mushroom mycelium. There is no precise point at which phase I should be translated to phase II; the more activity that takes place in one process, the less will be needed in the other, although too much activity in the first phase can lead to insufficient activity in the second. It is normal for the phase II process to last about 6 days. During this time the compost is either contained in the final growing container (shelves or trays [4]), or is processed in bulk. If in bulk, air is forced through the compost rather than around the container (5).

Traditionally, there is greater control of the environment in phase II than in phase I. At the beginning of phase II, the compost temperature is allowed to settle (often referred to as levelling) so that it is more or less uniform throughout. This may take 4 or 5 hours or more. Then the temperature is raised, either by allowing the fermentation process to generate heat, or by the introduction of steam. When the compost temperature reaches 60°C or just below, further temperature increase is prevented by the introduction of air (filtered). The temperature of 60°C is held for about 8–10 hours and this stage is often referred to as *the kill*. At this temperature, all the mesophilic organisms are killed, which includes all the pathogens and pests of the crop, but thermo-tolerant organisms remain. After the kill, the temperature of the compost is reduced to 48°C for the conditioning process. At 48°C, the thermo-tolerant fungi remaining, and in particular the fungus *Scytalidium* sp., grow quickly and colonize the compost. It is the biomass of these thermo-tolerant organisms that accumulates during the conditioning process that contributes to the compost's nutritional suitability for mushroom growth.

At the end of conditioning, which is usually about 4 days, the compost is cooled to near 25°C, so that it can be removed and spawned. It must be stable and virtually free from ammonia (5 ppm or less). Cooling may require large amounts of air and unless this air is filtered to remove pathogen or mushroom spores, there is a risk of contamination which may negate all the careful preparation in the production process.

4 Phase II in wooden trays. Note the compost is covered with paper to prevent air-borne contamination.

5 Bulk phase II compost being filled into the phase II room.

Compost smells

The very nature of the materials and the process can result in unpleasant smells, particularly at certain stages. The older systems of prewet and phase I, with no underfloor ventilation, often resulted in anaerobic conditions, especially in the centres of the piles of compost. When these piles were turned there was a considerable release of smells from the anaerobic areas. Under-stack ventilation and the use of bunkers have greatly reduced anaerobic smells. By completely enclosing the phase I and II systems and by the use of scrubbers and biofilters, the gases from the system can be practically odour-free, but this is a very costly process and not deemed affordable in most countries. However, the increasing pressure to reduce smells has resulted in a large increase in the use of bunkers with under-compost ventilation, which goes some way towards the reduction in smells. Unless biofilters are used, odour pollution is not prevented by the bunker system.

Compost analysis

There is no known ideal chemical composition for mushroom compost, and excellent crops can be produced within a range of analyses. Growers routinely have compost analyzed in order to monitor their own systems and to detect unplanned variability at an early stage. Both phase I and phase II composts are analyzed. Generally for phase I compost, the pH, water and nitrogen contents are obtained. pH is usually in excess of 8, water at about 75%, and nitrogen about 1.5–2.0% of the fresh weight. For phase II the figures are pH 7.2, water content at 68–72%, and nitrogen 2.5–2.7%. Generally the carbon to nitrogen ratio at spawning is about 15:1 to 18:1. When this ratio exceeds 20:1 there is an increased chance of weed moulds developing.

Mushroom compost: the selective medium

A selective medium is one which will grow a particular organism and no other. In this respect mushroom compost is not selective, but it has been preconditioned to be very good for the growth of mushroom mycelium. The process of compost production, in particular the phase II process, results in a medium that is not sterile but is loaded with thermo-tolerant organisms which are in a state of dormancy at spawning because of the suboptimal temperatures. This partial biological vacuum is filled by the introduction of large quantities of mushroom spawn. Extensive work has not been done on the range of other organisms that will grow well in phase II mushroom compost, but a number are well known as a result of accidental contamination at or before spawning. Recent work has shown that some of these moulds, while not inducing symptoms of disease on mushrooms, can have a commercially significant effect on their yields. Once colonized by mushroom mycelium, the compost is generally not vulnerable to infestation by other organisms.

Spawning, spawn-running, and phase III compost

Spawning

Once compost has completed the phase II process and has been cooled to 25°C, it is ready for spawning. At this stage, it is particularly important to be certain that the ammonia level in air in the compost is below 5 ppm. Spawn, mushroom mycelium growing on sterilized grains (rye or wheat and sometimes sorghum or millet), or less commonly on a grain-free medium, is thoroughly mixed into the compost by various mechanical means according to the growing system employed (**6**). A rate of spawn of 7–8 litres per tonne (or 0.5% by weight) of phase II compost is normal.

6 Spawning phase II compost. The spawn is pushed through the grid in order to break it up into single grains. Photograph by kind permission of Brian Oxley.

Spawn-running

Spawn-running is colonization of the compost from grain inoculum, and generally takes 13–18 days, according to the system used on the farm. The environmental conditions required for a successful spawn-run are, primarily, a compost temperature of 25°C, and high relative humidity to prevent the compost from drying. Carbon dioxide concentrations of 2%, or even higher, are beneficial, although care must be taken when entering houses with spawn-runs as very high concentrations of carbon dioxide are toxic to humans. This environment is achieved by recirculating air within the spawn-running room, the air being cooled if necessary. Spawn-running, like phase II composting, may take place in the final growing containers, or in bulk. Such spawn-running in bulk produces phase III compost, i.e. compost that is completely colonized by mushroom mycelium at the end of the process (*see* below and p. 20).

Generally, spawn manufactured by specialist producers, is free from pests and pathogens. When containers of spawn are opened on a commercial farm, the spawn is vulnerable to contamination.

The introduction of pests and diseases during spawn-running can be very damaging. As a generalization, the earlier that contamination takes place, the worse the effect on the crop. It is therefore essential that, during the spawn-run process, air used to control the compost temperature is always filtered (7), and the hygiene in and around the spawn-running area is particularly strict.

Phase III compost

It is now common practice in a number of countries to prepare fully colonized compost in bulk. This is done in large tunnels which may be the same as those used for phase II, or they may be dedicated to the phase III process. Temperatures during spawn-run are controlled using filtered cool air which is blown up through the compost. In this way the optimum temperature of 25°C can be maintained, and the compost is fully colonized in 16–18 days. It is then removed from the tunnel and transported in bulk to shelves or to machinery where it can be put into trays, bags or blocks (8). Controlling oxygen levels

7 A filter housing with a blue pre-filter showing.

throughout the bulk spawn-running process is very important, and it is generally believed that the oxygen level during phase III should not fall below 16%. Very strict hygiene is required with phase III production, as pathogens, weed moulds, mushroom mycelial fragments, and possibly mushroom spores from a nearby crop, can initiate diseases.

Specialist phase III producers sell compost, eliminating the need for a farm to have its own compost producing facilities.

Supplementation

Compost is often nutritionally supplemented, normally at the time of spawning. Some systems of growing allow this to be done with advantage, just before casing.

While the mechanisms of this nutritional boost are not fully understood, the beneficial effects on yield levels are often thought to be at least cost-effective, and can be as high as 20%. The supplement, a product with a high protein content often soya-based, treated by heat or formaldehyde to give a slow-release preparation, is mixed into the compost at the same time and in the same way as the spawn. The technique is not used where the control of high compost temperatures is known to be a problem, or where adequate mixing is difficult, or where the compost is of doubtful quality.

Casing

To promote mushroom production, it is necessary to add a relatively biologically and nutritionally inert surface layer to the colonized compost (**9**). This 'casing' layer was traditionally a mixture of baled peat and limestone. Types of peat and chalk vary with the country, although now in many countries a mixture of well humified black peat (from the deeper levels of the peat bog) and a by-product of the sugar beet industry called sugar beet lime is used. This mixture remains open and continues to drain effectively throughout cropping, allowing more water to be used for the control of pinning and cropping.

8 Blocks of phase III compost.

9 Casing application to trays.

The casing layer is applied 4–5 cm deep. It must have a neutral or alkaline pH. In addition to stimulating fruiting, it provides anchorage for the mushrooms, and also the water-holding reserves essential for high yields. Casing material is easily contaminated and if this occurs it can result in serious outbreaks of pests and diseases. Casing ingredients can be heat-treated to remove pests and pathogens. They must not be sterilized but heated to a temperature that is sufficient to kill harmful organisms, while retaining the bacteria that are important for cropping. Steam–air mixtures have been used for this purpose and when in equal proportions the mixture has a maximum temperature of 80°C. Ideally a temperature of 60°C should be used, and the casing heated for 30 minutes.

Spawned casing

The use of mushroom inoculum in the casing layer is now established practice. There are two ways of doing this.

Casing inoculum (CI) can be purchased from spawn suppliers; this is mushroom mycelium growing on very small fragments of a low-nutrient medium such as peat or mica. Ordinary spawn is unsuitable for this use. The inoculum is mixed with the casing when it is spread, or it is evenly distributed onto the surface of the casing after application and then rotovated or raked in.

An alternative is spawn-run compost mixed into the casing in the same way as for the casing inoculum. This process is known as cacing (compost added to casing). The evenness and speed of colonization (case-running) is greatly enhanced by the use of casing inoculum. While the advantages of this technique are considerable in terms of the time and evenness of cropping, there are serious implications for pathogen and pest development particularly if spawn-run compost is used. The success of this method depends entirely upon the selection of compost, free of pests, pathogens and moulds.

Ruffling

This is an alternative to the use of casing spawn, or cacing. The casing is either deeply rotovated immediately after application and some of the colonized compost from the top layers of the compost is mixed into it, or rotovation is done when the mycelium has grown from the compost into the lower layer of the casing, and this mixes the mycelium with the casing (**10**). These techniques can have the same effect as the use of casing inoculum, although they are less reliable as they are more difficult to regulate.

Ruffling the casing after a number of days of mycelial growth may also make the first flush more even and help to regulate its timing. It can be a very effective way of spreading disease.

Allergies

Micro-organisms, particularly actinomycetes, are produced in large numbers during composting. 'Firefang,' a white growth in the compost seen at the end of phase II, is a prolific growth of these. During handling compost, at any stage, many organisms become airborne. Workers constantly exposed to these organisms may develop mushroom growers lung. It is important therefore that those working

10 Equipment used to ruffle a shelf crop. The tines mix spawn-run compost with the casing. Photograph by kind permission of Brian Oxley.

with phase II and phase III compost wear an aspirated helmet.

During cropping there are often large numbers of mushroom spores in the air. While there is a risk that some pickers may develop allergies to these the evidence suggests that this risk is very small.

Cropping

Once the mycelium has reached the surface of the casing, the crop is induced to fruit. This is done by reducing the air temperature (to 16–18°C) over a number of days (3–5), and also by reducing the carbon dioxide concentration in the air (to about 1000 ppm). It is the casing temperature that is critical, as it is here that the mushroom fruiting initials develop. Fruiting occurs in well defined flushes or breaks, the first beginning about 17 days after casing, and continues at roughly weekly intervals (**11**). In commercial practice three flushes are picked, and then the crop is removed to make room for the next crop. There is an increasing trend to take only two flushes.

About 6–13 crops can be taken from a cropping house every year, depending upon the system used. The first two flushes produce the most mushrooms and are about equal in size. The third flush produces about half the quantity of mushrooms of each of the first two. If subsequent flushes are taken they produce progressively less, and for this reason are usually considered to be uneconomic.

In order to prevent epidemics, and also to keep pests and pathogens at a permanently low level, especially in organic crops, only two flushes may be taken.

Access to the crop by personnel is greatly increased during cropping. Casing is watered at regular intervals from the time it is applied up to the point just before mushrooms are initiated (**12**). Watering is started again after the clearance of the first flush, and often between the second and third flushes. Most water is applied before the first flush, and as much as 25 litres per m^2 can be used at this stage.

Pickers move in and out of the houses to harvest crops. Any pests or pathogens that occur usually increase in incidence as the crop ages. Pests and pathogens spread within crops and also between crops, particularly from the oldest to the youngest. Pest and disease identification is crucial if levels are to be controlled within reasonable bounds, and effective methods of crop termination, compost emptying, and disposal are essential (*see* Chapter 3).

11 First flush mushrooms ready for harvest.

12 Watering mushrooms with a watering tree.

Pests and pathogens

There are clearly some stages in mushroom production that are more vulnerable to the entry and establishment of pests and pathogens than others. These have been mentioned very briefly in this chapter, but are discussed in more detail in the following chapters. An overall picture of the stages in crop production (**13**) shows the critical times in relation to pests and pathogen risk.

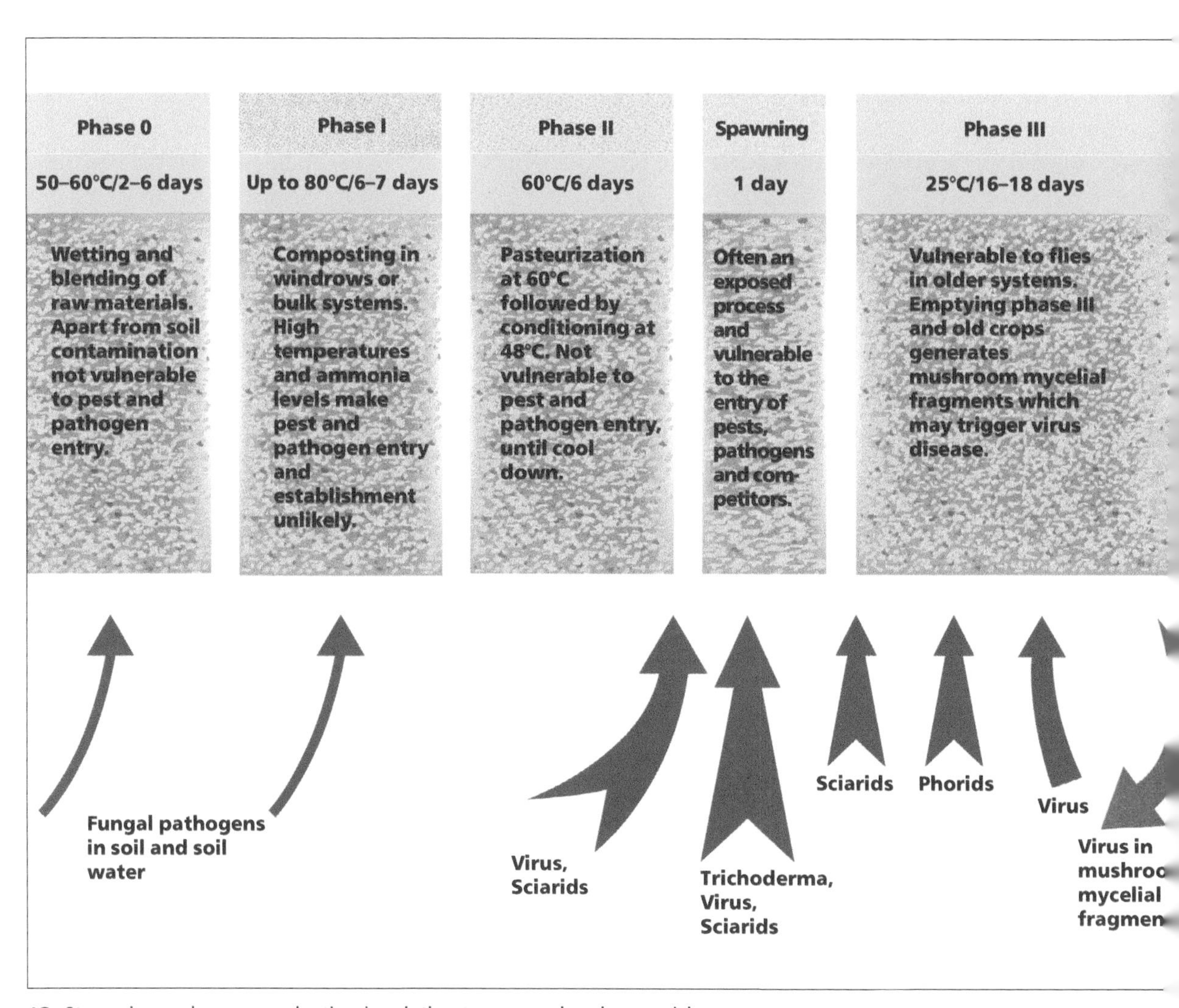

13 Stages in mushroom production in relation to pest and pathogen risk.

13

Casing	Case-running and cropping		Crop termination
1 day	25–18°C/31–38 days		65–70°C for 12 hours
An exposed process vulnerable to the entry of fungal pathogens and mushroom spores and mushroom mycelial fragments.	Vulnerable to fungal and viral pathogens during case running.	By the end of cropping large populations of pests and pathogens are possible that may endanger other crops, concurrent and future.	Major pest and pathogen elimination

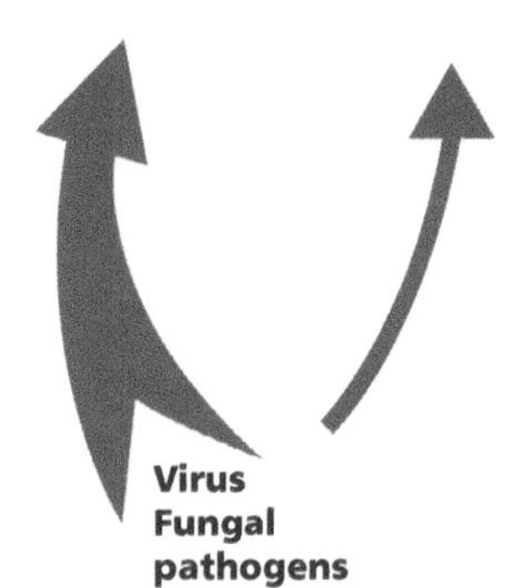

Mushroom spores with virus

Virus
Fungal pathogens

Primary source of pathogens and pests

Mushroom growing systems

Growing systems have evolved as mushroom production has developed. No one system is suitable for all purposes. Systems as diverse as shelf and bag growing can be equally profitable. Originally, mushrooms were grown in beds on the floor of the growing area (**14**). This system of culture is now infrequent, although it is used to a limited extent in cave growing. Containers of various sorts evolved from the bed system. Although shelves, as containers for compost, have been used for many years especially in the USA, modern shelves, as developed by the Dutch industry, are probably the most effective system (**15**). The tray or box system of growing is widespread and still thrives in many countries (**16**). The Irish industry, which uses plastic bags and more recently blocks, is also very effective and far less capital-intensive (**17**). A system of growing in which the compost was put into deep troughs has largely disappeared.

On some farms, all the stages of crop production from the beginning of phase II to the end of cropping are done in one room. Such a system is known as single zone, and is common in the USA. More usual in Europe, and elsewhere, is the two zone system in which phase II is done in specialized rooms, and spawn-running and cropping in others. In the three zone system there is physical separation of each stage, i.e. phase II, spawn-running and cropping are all done in separate facilities. A four zone system, in which crops are case-run and aired in specialized facilities and then finally cropped in others, sometimes on another farm, is becoming more common.

14 Mushrooms as they were grown (c. 1930) on the floor of the house. Although the compost was similar to that used today the casing was soil.

15 Shelf system of growing. The shelves may be up to seven high depending upon the structure.

16 Tray growing with the trays stacked four high.

17 Extra crop is being produced by having a double layer of bags.

Tray systems

Trays vary considerably in size but are mostly in the range of 0.9–1.2 m by 1.2–2.4 m and are between 20 and 23 cm deep. They are generally wooden, although metal trays are also used. The trays are moved by fork lift from one room to another, and the processes of tray filling, spawning and casing are done on mechanized tray-handling lines (**18**). In the cropping houses, the trays may be stacked three, four or five high. Compost may complete its phase II process in trays, and at spawning the compost is emptied from the trays, before spawned material is refilled back into the trays. Bulk phase II compost and, in a few cases phase III compost, may also be used in trays. Where this practice is adopted, the trays have not passed through the pasteurization process and are, therefore, more likely to be contaminated after cook-out, unless they have been well protected.

It is now common practice to line trays with thin polythene, which is replaced for every crop. This not only reduces the risk of carry-over of pests and pathogens, but also makes tray cleaning easier.

A further development of the tray system is the growth of the crop up to the point of pin-head mushrooms, followed by moving the trays to the cropping houses. This is sometimes done on the farm when the trays are held at the pin-head stage in a so-called holding room or set back room before being moved to the cropping room. In a more recent development in some parts of Europe, the trays (often metal) of compost may be developed to the pin-head stage in one country, and then moved and cropped in another. This has been called the pinned tray or the phase IV system.

18 A tray line in operation at casing.

Shelf systems

Phase I or phase II compost, or increasingly phase III compost, is filled onto shelves, up to six high. Originally shelves were made of wood which could be easily removed. Such shelves are commonly used in the USA. Elsewhere, metal shelves have largely replaced wooden ones and the 'Dutch system' of shelf growing is probably the most highly evolved in the world. All shelf systems require special equipment with which to fill the shelves and to apply the casing. It is particularly important that at crop termination the crops can be cooked-out *in situ*, as the movement of compost from shelves into trailers can create dust and debris which consists of spores, fragments of mycelium and pests which may then be recirculated around the farm. Phase III shelf growing can be the most productive of all the mushroom systems. Mechanical harvesting is possible with the metal shelf system, and is often used if the mushrooms are for canning.

Bag and block systems

Plastic bags, which vary in height and width and have a bed surface area of 0.1–0.2 m^2, are usually filled with approximately 20 kg of spawned phase II compost, or more recently phase III compost. The system has been used very successfully in Ireland where, as a result of the socio-agricultural structure and commercial composters/marketers, it has formed a very effective industry. The bags of compost may be tiered on racks, but in the simplest and perhaps the best system, are placed on one level on the floor of the cropping house.

A modern variation on the system, and one that is increasing in popularity, is the use of phase II or III compost compressed into blocks (about 20 kg) and wrapped in plastic. The blocks are then fitted into existing trays or shelves, or are used in a distinct system on specially made racks, tiered two, three, or four high (**19**). The spawn colonizes the phase II blocks before the plastic on the top of the block is removed and the casing applied. With both bags and blocks, cropping is well separated from the phase II composting, and this generally reduces the incidence of pests and pathogens, providing the phase II and III compost is hygienically prepared.

19 Phase III blocks loaded into shelves.

Deep trough system

In this system, bulk phase II composting, spawn-running and cropping are done in a deep trough, under which runs a plenum (ventilation system) allowing air to be forced or sucked through the compost in a manner similar to that in bulk processing tunnels. The system has merit for the very small grower as it requires a minimum of equipment and can be successful with small quantities of compost. It is actually however, a very sophisticated system, and may have failed to become established for this reason.

Buildings

Types of mushroom-production buildings are very variable, but all must enable the grower to achieve good environmental control. Almost every type of building imaginable is employed, as well as underground quarry galleries and railway tunnels.

Major requirements are very good insulation, an effective heating and ventilation system, the ability to clean thoroughly, and a capacity to withstand the heat of the cook-out many times without distortion. Traditionally the commonest structure used in the USA is the 'standard house' constructed of cement or cinder blocks, with a wooden framework inside arranged to make two tiers of beds, each consisting of six shelves. The boards of the shelves are loose so that they can be lifted to observe the growth of the mycelium in the compost. The beds are filled from the top of the house, and as each layer is filled the base boards of the next layer are put into place. Cropping area can be increased by building two such houses together to form an 'American double' (**20**). Originally such houses were built on a slope so that the compost could be wheeled directly into the top layer of the house. Now mechanical loading equipment is used.

Insulated plastic tunnels are increasingly common in many countries and are used not only as growing rooms, but also for the preparation of phase II compost (**21**). They can be made to meet various requirements while being relatively cheap to build. Many older farms in Europe use buildings with portal-frame and block-walls of brick or stone (**22**). Farms built on the Dutch design are factory-like structures often of aluminium-coated insulated panels (**23**). Even where purpose-built houses are used, arrangement of these on a farm may have considerable bearing on pest and disease control, depending upon the overall farm design. Where farm buildings and other types of buildings and covered spaces have been converted for mushroom production, there are often major limitations which make an apparently cheap structure uneconomic.

20 American doubles, the type of cropping houses very commonly used in the USA.

21 Plastic covered insulated tunnels, the cheapest form of structure for producing the crop.

22 Portal-frame and block-wall cropping houses.

23 A modern 'factory like' Dutch designed mushroom farm. Photograph by kind permission of Steve Newton.

CHAPTER 2

Disorders: Symptoms, Causes, and Identification

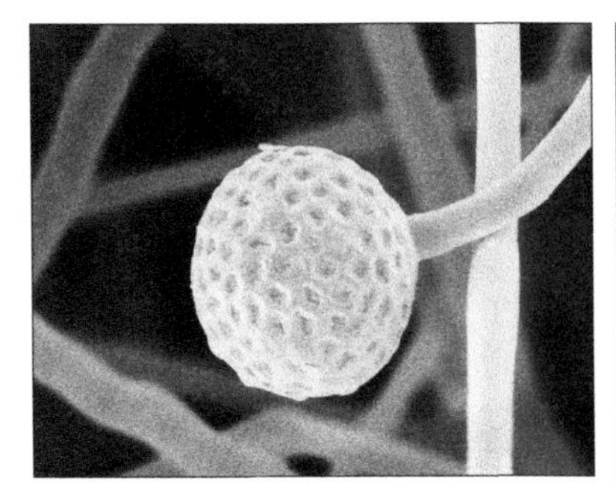

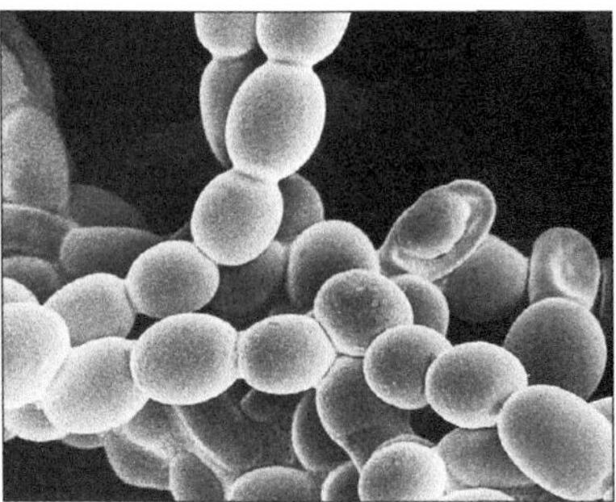

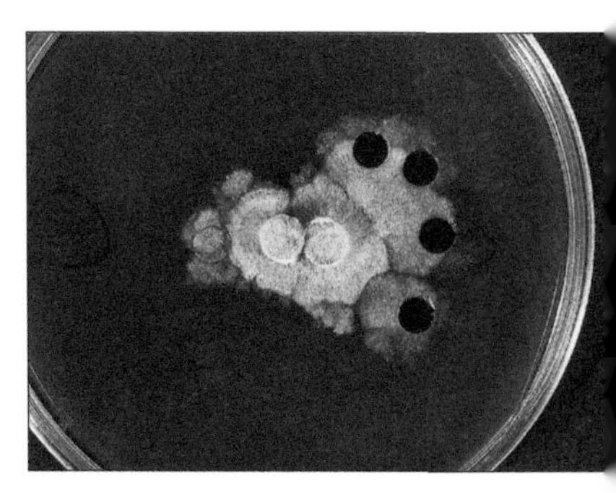

Introduction

The success of mushroom culture is dependent upon many interacting factors and it is not unusual for crops to achieve less than the maximum yield. In addition, some of the mushrooms from each crop are less than perfect in shape or colour and the crop value is consequently reduced. Biotic factors, the various pests and pathogens, often produce the most obvious defects, although some abiotic factors, such as abnormal environmental conditions and the presence of toxic chemicals, can be equally damaging. In this book we are concerned chiefly with the biotic factors and the symptoms they cause, although abiotic disorders, often with no known cause are also described.

By definition a disorder, whatever the cause, is a deviation, however small, from the normal. But from the practical viewpoint, a disorder produces recognizable symptoms that often result in a reduction in yield and/or quality.

Names of disorders

Choosing a name after a predominant symptom is often the preferred choice, but with pests it is more usual to use the name of the pest. Where there are multiple symptoms, or two or more disorders with similar symptoms, a common name based on a symptom may be difficult to find. The virus diseases are good examples of diseases with multiple symptoms and the two bubble diseases share a common symptom. Sometimes the Latin name of the causal organisms is used as the common name, e.g. *Verticillium* and *Mycogone*. Common names can be confusing and green mould and virus X are good examples of these as they have both been used for more than one disease.

Causes of disorders

The expression of the symptoms of disorders in mushroom culture depends upon two major factors:

- The cause of the disorder.
- The stage of development of the crop at the time the disorder develops.

Growers are generally most concerned about biotic factors as it is the persistence of these on the farm that affects profitability. Some of the biotic causes of disorders are easy to identify, but symptoms may be similar, whatever the cause, and may result from an interaction of a number of factors, both biotic and abiotic. Accurate identification wherever possible is essential, so that the most effective means of control can be used. Abiotic factors are generally more sporadic in occurrence and are frequently more difficult to explain. Occasionally they can cause disorders that are as devastating as those caused by biotic factors, e.g. persistently high carbon dioxide levels which result in delayed, underpinned or deeply pinned crops; or periodic toxic pollution from nearby industry producing discoloured mushrooms.

Symptoms in relation to crop development

The mushroom crop has well defined stages in its development, and symptom expression is often dependent upon the stage the crop has reached when the problem begins. The first critical stage is the introduction of healthy mycelium into the compost, which, in normal circumstances, is followed by extensive mycelial growth until the whole of the compost is colonized. In the next stage, the production of mushrooms is induced by the application of a casing layer on top of the colonized compost. Soon after the casing layer has been applied, the mushroom mycelium undergoes a sequence of physiological and morphological changes, ultimately resulting in the growth of the fruiting bodies (sporophores) of the fungus, which we recognize as mushrooms.

Factors affecting mycelial growth often result in non-cropping areas or poorly colonized compost. At a microscopic level, the first indication of mushroom formation is the aggregation of the mushroom mycelium on or near the surface of the casing. The stimuli which induce the production of mushroom initials may on occasions be super-efficient and vast numbers are formed. Differentiation is followed by a stage of rapid cell enlargement which, when uninhibited, results in good quality mushrooms. Once mushrooms have been initiated, both biotic and abiotic factors can result in symptoms.

It is easy to see that any disturbance in the development of the normal crop will result in symptoms that are recognizable to the grower. For instance, poor or patchy mycelial growth will become apparent at the time of applying the casing. Disorders which start after mushrooms have been initiated but not fully formed will result in distortion. Surface blemishes appear on the caps of partly or fully developed mushrooms. Very often the symptoms are diagnostic, and give a good indication of the time that the problem began, which in turn can suggest the cause.

Biotic factors responsible for disorders

The most commonly encountered biotic causes of disorders are insect pests, mites, nematodes, pathogenic fungi, antagonistic fungi, pathogenic bacteria, and viruses.

Insect pests

Many disorders are caused by species of Diptera (flies). Flies are attracted to the mushroom crop and their larvae may feed directly on the mycelium, swarm over the mushrooms, or tunnel into the developing, or developed, mushrooms. The symptoms resulting from fly attack vary from a reduction in yield, due to direct or indirect effects on mycelium, to discolouration and damage to the mushrooms resulting from direct attack. There may also be spoilage, as a result of flies being present in enclosed pre-packs. Tissues that have been physically damaged by fly larvae often become colonized by bacteria that cause soft rot, thereby accentuating the problem.

Mites

Like fly larvae, some mites may feed on mushroom mycelium and on developed mushrooms, where they can cause surface discolouration and reduce yields. Others may live on fungi other than *Agaricus bisporus* (e.g. the antagonistic fungi, see below) found in mushroom culture, some feeding on decaying organic material, and some on nematodes. They can be symptomatic of poor compost preparation. When mites are numerous they can be a source of irritation to pickers, thus reducing harvesting efficiency.

Nematodes

The most destructive nematodes (parasitic) are those that feed on mushroom mycelium, resulting in bare patches or sunken areas of bed which can reduce crop yield significantly. Those that live entirely on organic matter, such as mushroom casing (saprophytic), are frequently found in mushroom crops and can reduce yield, but only in very specific circumstances. Nematodes are very small and only visible when they aggregate into clusters on the casing surface. Large numbers of saprophytic nematodes are often symptomatic of problems in compost preparation.

Parasitic fungi

Various fungi are known to be parasites of the cultivated mushroom. Mycological identification of these fungi is generally based on their spores or spore-producing structures, but the diseases are recognized by the symptoms shown by affected crops. Most of the common fungal pathogens attack mushrooms and not the mushroom mycelium. Generally the earlier the attack the greater the yield loss and the more distorted the ultimate mushrooms; for example, very large distorted mushrooms result when mushroom initials are attacked by *Mycogone perniciosa*. Some fungi (for example *Diehliomyces microsporus* and *Trichoderma aggressivum*) attack mycelium of the crop.

Symptoms of fungal attack can vary from decay (cap spotting and stalk rot), to severe distortion (malformed mushrooms), to complete loss of yield.

Antagonistic fungi

The relationships between mushrooms and other fungi that can occur in mushroom crops are only imperfectly understood. Many of these fungi become established at or soon after spawning, because the physical and chemical environment of substandard compost is favourable for their growth. Some moulds are able to grow in well prepared compost and then compete with mushroom mycelium for nutrients; others are antagonistic, and once established, prevent the mushroom mycelium from growing into affected compost. In both cases, the end result is yield reduction.

It is also likely that some moulds produce toxins and some of these may be volatile. Toxins can induce distortion in developing mushrooms.

Moulds can sometimes be recognizable in the compost or on the casing surface, by the coloured spores or mycelium they produce, and for this reason they have frequently been given descriptive names such as Olive green mould, Plaster mould and Lipstick mould. In modern methods of production they are much less obvious, but in some ways, just as dangerous.

Pathogenic bacteria

Some bacteria have a vital role in the successful production of mushrooms, but others can cause serious disorders. World-wide, the most common and most investigated bacterial disorder is Bacterial blotch (*Pseudomonas tolaasii*), which discolours and sometimes disfigures the developing or more frequently the mature mushrooms, even after marketing.

There is some evidence that pathogenic bacteria can be present within apparently healthy mycelium with no obvious effect on mycelial growth. Symptoms may then develop when mushrooms are produced. For instance, bacteria (or similar organisms) that are present within the mycelium are the causes of such diseases as Drippy gill and Mummy disease. Bacteria are known to cause distortion, discolouration and decay; they may also be responsible for delayed mushroom production. After harvest, they are a major cause of mushroom browning.

Viruses

Many fungi are known to contain particles which are very like those of plant viruses. There are relatively few instances where the presence of these is associated with disease symptoms, *Agaricus bisporus* being one of the few. There are no known vectors of mushroom viruses.

Mushroom viruses are transmitted in mycelium and also in mushroom spores. Exhaustive work on the significance of the various types of virus particle found in mushrooms has yet to be completed. The precise effects of these, individually and in combination, is still not known, and there is therefore an element of uncertainty concerning the symptoms produced by specific viruses in mushrooms. There is increasing evidence that the traditional virus disease of the crop (La France virus disease) is caused by only one of the various viruses frequently found in apparently healthy crops. Whether the other viruses contribute to symptom production in any way is not known. A number of different dsRNAs (mushroom viruses are predominantly double-stranded ribo-nucleic acid), apparently not contained within a protein coat, have been associated with the new Mushroom virus X disease.

Mushroom virus diseases cause considerable reductions in yield. Symptoms that have been attributed to virus diseases include growth abnormalities of mycelium, discolouration and distortion of mushrooms such as browning of caps and elongation of the stalks. Early opening of caps, crop delay, delayed pinning development often in specific areas of the crop, and die-back of the mycelium, may also be symptoms of virus diseases. Most of these symptoms can also be caused by other factors.

Viroids, mycoplasmas, and rickettsias

These three groups are known to cause diseases in plants and animals, but have not so far been found to cause diseases in mushrooms. They are all difficult to study and easy to overlook. It is possible that, individually, they may be playing an important role in some of the diseases at present attributed to other causes, e.g. viroids in virus diseases and rickettsias or mycoplasmas in bacterial diseases such as mummy disease.

Abiotic factors responsible for disorders

The most common abiotic causes of disorders include suboptimal environmental conditions in the atmosphere, poorly produced compost, poor casing, the presence of toxic substances particularly in casing, and genetic abnormalities of spawn. Either alone or in combination, these factors may result in poor mycelial growth and poor cropping or distortion of the developing or developed mushrooms. Symptoms produced include hard gill, rosecomb, and mass pinning. Examples of many of the abiotic disorders of the crop are given in Chapter 9.

Abiotic causes of mushroom disorders are often difficult to identify. The cause of the symptom may be transitory, and by the time the symptom has appeared the cause has disappeared. A good example of such a disorder is the failure of the first flush to develop normally. This is possibly the result of adverse conditions at the time of pin-head production, which does not become apparent until the first flush begins to develop. Even when historical evidence is available which clearly points to particular abnormal factors, it is often very difficult to reproduce the symptoms by applying these factors to an experimental crop.

Identification of disorders

There are a number of stages in the identification of disorders in mushroom crops. The first stage is usually an on-the-spot investigation and the gathering of information about the disorder. In some cases this is sufficient to give a clear indication of the cause, but in others it may be necessary to take samples for laboratory examination. Both chemical and biological analysis may be required.

Gathering information

Answers to the following questions will give information which may immediately lead to the cause or at least enable the examination to proceed:

1. Are the symptoms like any described in the literature?
2. When was the disorder first seen on the farm?
3. What was the distribution of the disorder both on the farm and in affected crops?
4. Did the disorder occur irrespective of compost or casing source or type, or was it related to a specific material?
5. Have laboratory tests been done on affected mushrooms and/or compost?
6. Was the disorder confined to one spawn type or were different spawns affected?
7. Was the affected crop treated differently from the normal ones?
8. Have the environmental settings been achieved for the affected crop?
9. Have pesticides been used on the crop and was the rate of application correct?

Many other factors involved in mushroom production can be considered in the same way, for instance: environmental changes both external and internal; methods used in the production of the compost and casing; the general hygiene practices on the farm; and methods of application of pesticides. If recent changes have been made to any of these processes they should be considered to be especially suspect. The results of laboratory tests and analyses of ingredients can supplement this information.

Frequently a methodical approach will result in identification, or at least suggest a probable cause. However, in some instances an obvious answer cannot be found, especially when the study is confined to a single occurrence on one farm. Disorders that are difficult to identify are sometimes resolved when the disorder is more widespread, and an analysis from a number of farms shows that there are some factors in common to them all.

When disorders have a biotic cause, identification is generally not too difficult. The speed of identification depends upon the organism involved and the ease with which it can be seen. Pathogens new to the crop must be used in inoculation experiments in order to reproduce the symptoms, before the cause can be positively confirmed.

Abiotic causes are much more difficult, and an investigation using a methodical approach taking into account all possible factors, is usually the only one possible. Sometimes hypotheses can be tested by applying factors to experimental crops. Reproduction of the symptoms is considered to be positive proof, but negative results do not necessarily prove that the tested factors were not involved. If the cause is an interaction between various factors and the developmental stage of the crop, it is usually very difficult to reproduce all the conditions.

Taking samples for laboratory examination

Compost/casing samples

It is very important that compost or casing samples are representative of the batch if they are taken for a routine test, or of the problem if they are being checked as a possible cause. Twenty subsamples taken at random should give a representative sample. The larger the number of subsamples the more representative the final sample. The size of the subsamples can vary according to the total amount of material to be sampled. The subsamples should be thoroughly mixed in a large polythene bag and a final sample of approximately 1 kg taken for laboratory analysis. An accurate system of labelling is essential so that the results can be related to the materials sent.

Mushrooms

Those showing characteristic symptoms are carefully selected using gloves and taking the precautions associated with disease removal (*see* p. 41). If possible, mushrooms for biological analysis should not be contaminated with casing. Each mushroom should be carefully wrapped in paper (newspaper is satisfactory) and packed in a box with additional paper to prevent them moving and being damaged in transit (**24**).

Moulds

Those growing on casing are difficult to sample without destroying the symptoms. The ideal is to take a small square of casing showing the typical mould, and carefully remove it from the bed into a rigid container of the same size. A lid should then be placed on the container which should then be tightly packed so that the sample cannot move.

Moulds in compost are sampled by taking the affected compost and packing it in a polythene bag.

24 Mushrooms sent to the laboratory for identification of pathogens. (**a**) The correct way, note that each affected mushroom has been cut with no casing material and carefully covered in paper. (**b**) Incorrectly harvested and packed, the affected mushrooms have become totally covered in casing and are impossible to examine.

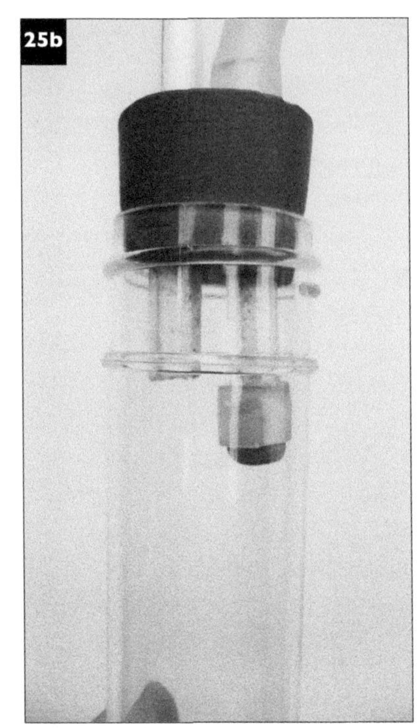

25 A pooter (also known as an aspirator), used to collect flies and other insects for examination. The long tube is pointed at the insect, and by breathing in the insect is sucked into the tube (**a**). The insect remains in the tube because of the filter (**b**). A cork or cap can then be placed on the tube and the insect can then be examined in the laboratory. Thanks to Dr Mike Copland for demonstrating the technique.

Identification

Insects, mites, and nematodes

These are all macroscopic organisms varying in size from 1 to 5 mm. When associated with a disorder, they are usually present in large numbers and are not difficult to see. Nematodes are the most difficult, although saprophytic nematodes sometimes swarm and can be seen in the compost or on the casing surface. Shining a strong light parallel with the casing surface can reveal small swarms of nematodes. Precise identification depends upon a number of morphological features, together with behaviour patterns. In order to study the organisms in detail, it is necessary to examine them closely. For this purpose flies and mites are caught using a pooter (aspirator) (25). A low magnification microscope (up to 1003) or a good hand lens is also required. The morphological and other distinguishing features of these organisms are described in Chapter 8. Recognizing nematodes with the aid of the microscope is easy, but distinguishing between the saprophytic and parasitic types is a job for the expert.

Fungi

Most fungi, like the mushroom, produce vegetative mycelium and spores. It is the spores that are most useful in making precise identifications. Frequently, affected mushrooms are colonized by a fungal pathogen, and identification is then done by microscopic examination of a small sample. Pathogens such as species of *Verticillium*, *Mycogone*, and *Cladobotryum*, and many of the moulds, produce characteristic spores (**26**).
Not all fungi produce spores readily; where sporulation is difficult it can sometimes be induced by growing the fungus on another medium (often agar supplemented with nutrients).

It is usual to culture samples of compost on agar so that moulds can be isolated and identified. Where quantification is required, a water extract is made (using a Stomacher) which is diluted sufficiently to allow the colonies of fungi that develop on the agar to be counted. The numbers, generally the higher the more significant, can be very important when identifying the cause and potential of a compost problem.

26 Spore shapes and forms of some fungi associated with mushroom cultivation. (**a**) *Arthrobotrys*; (**b**) *Doratomyces*; (**c**) *Chrysosporium*; (**d**) *Geotrichum*; (**e**) *Mortierella*; (**f**) *Scopulariopsis*; (**g**) *Sporendomena*; (**h**) *Trichoderma*. By kind permission of Albert Eicker.

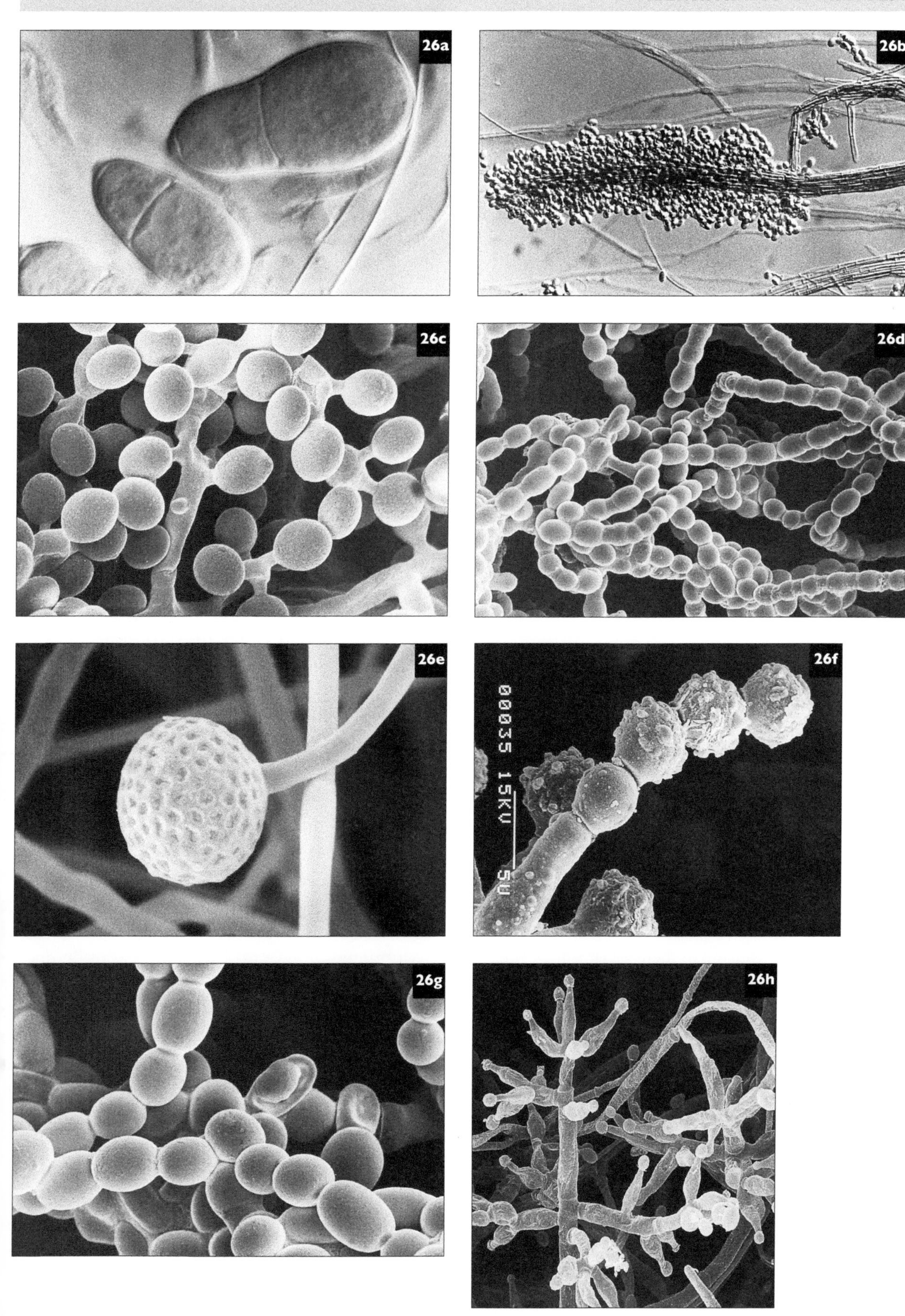
26a
26b
26c
26d
26e
26f
00035 15KU 5U
26g
26h

When more than one species or strain of a fungus could be involved, it may be necessary to make very precise identifications. For instance, there are various species of *Trichoderma*, and strains of *T. aggressivum* (syn. *T. harzianum)*, that cause different symptoms. Differences between species may depend upon clearly defined morphological features, such as spore shape and size, but within the species, strain differences may not be morphologically clear. For this reason, the various strains of *T. aggressivum* are precisely identified using DNA technology, or by comparing their growth rate, and speed of sporulation, in standardized conditions.

Also involved in fungus identification may be the characterization of the pathogen for its sensitivity to the commonly used fungicides. For this purpose, the pathogen is isolated from affected mushroom tissue and grown on nutrient agar in sterile conditions. Subcultures of the isolate are made onto nutrient agar into which different concentrations of the test fungicide have been included. During the incubation period, measurements of the growth of the pathogen are made at frequent intervals, and growth graphed against fungicide concentrations (**27**). The result of the resistance test is expressed as an ED_{50} value. This represents the concentration of fungicide at which the growth of the test fungus is reduced by 50%. Further details on fungicide resistance are described in Chapter 3 and examples of changes in the sensitivity of pathogens to fungicides in Chapter 4.

Bacteria

Bacteria are unicellular organisms, and those that are present in mushroom crops do not form spores. There is some variation in their morphology but not enough to enable them to be identified using this factor alone. Bacterial identification is dependent upon a series of chemical tests in which materials are added to agar media, and the ability of the bacterium to utilize these and to grow on the media is measured. In the case of *Pseudomonas tolaasii*, the cause of Bacterial blotch, precise identification is achieved by the white line test. This is a very specific interaction with a related bacterium, which when together with *Ps. tolaasii* on the same agar plate, results in the production of a white line of precipitation between the two bacterial colonies (**28**). Recently DNA technology and analysis of fatty acid profiles have added to the speed and precision of bacterial identification.

Viruses

Virus particles are very small and cannot be seen with a light microscope. Unlike fungi and bacteria, they do not grow on artificial media. Fortunately, modern techniques have greatly aided the study and identification of viruses, including those found in mushrooms.

The first described virus disease, La France disease, was initially identified by comparing the growth rate of mushroom mycelium from an affected mushroom, with that of the parent spawn. When virus is present at high concentrations, the growth rate of the affected mycelium is less than a half of that of the healthy strain. Because all strains may not grow at the same speed, it is important to make comparisons with a known healthy culture of the same strain.

The infectious nature of a La France virus affected culture can be readily demonstrated on agar by anastomosis (hyphal fusion) with a healthy culture (**29**). It can be shown that, within days of anastomosis occurring, the previously healthy culture grows much more slowly, and at a rate comparable to that of the diseased culture, demonstrating the presence of the virus and its transfer.

The electron microscope (EM) has been an important tool in the identification of La France disease. Affected mushrooms are squeezed and a small volume of the juice is put on a grid for examination. If present, virus particles can be seen in the preparation. A problem with this technique is the uneven distribution of the particles in affected mushroom tissue, and sometimes their low numbers. For this reason it is important to examine the juice from more than one mushroom, and to look at a number of fields on each grid. Usually 10 mushrooms are taken from an affected crop, and most consistent results have been obtained by sampling the third flush, taking apparently normal mushrooms from around affected patches.

The precision and consistency of results by electron microscopy is improved with a virus specific antiserum which is used to coat the microscope grid (immunosorbent electron microscopy or ISEM). When

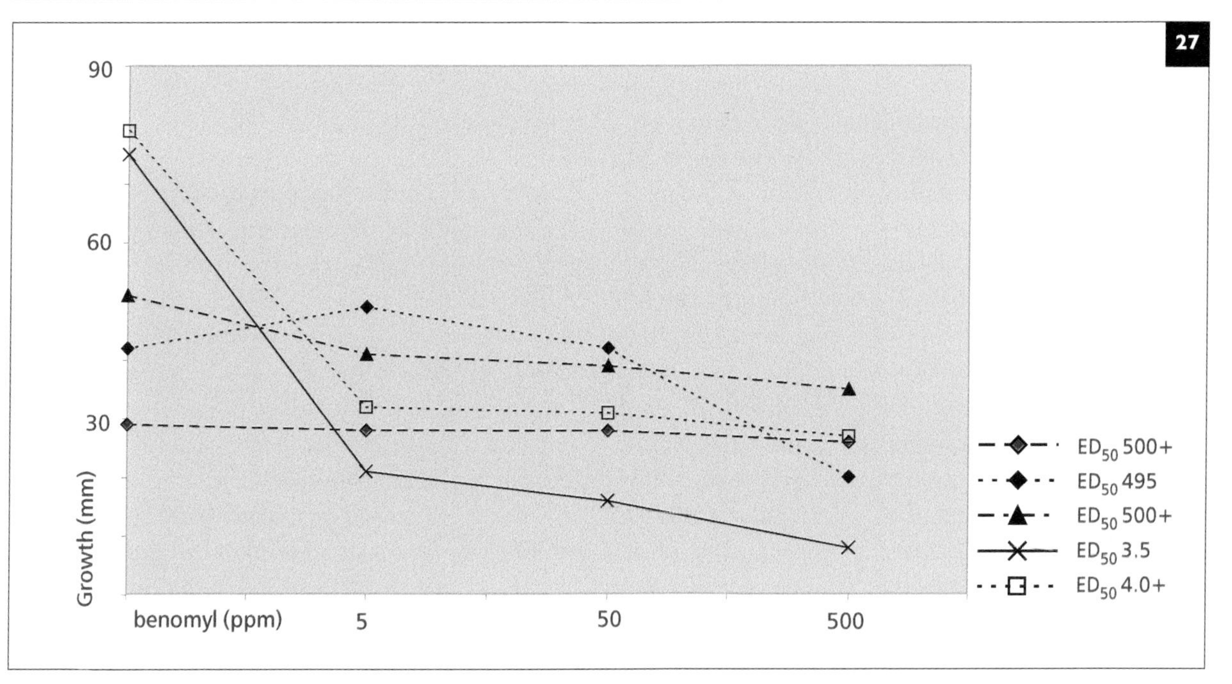

27 Fungicides resistance graph showing growth rates of five isolates of *Verticillium fungicola* on benomyl amended agar (after Fletcher and Yarham 1976, *Annals of Applied Biology*, 84, 343–353).

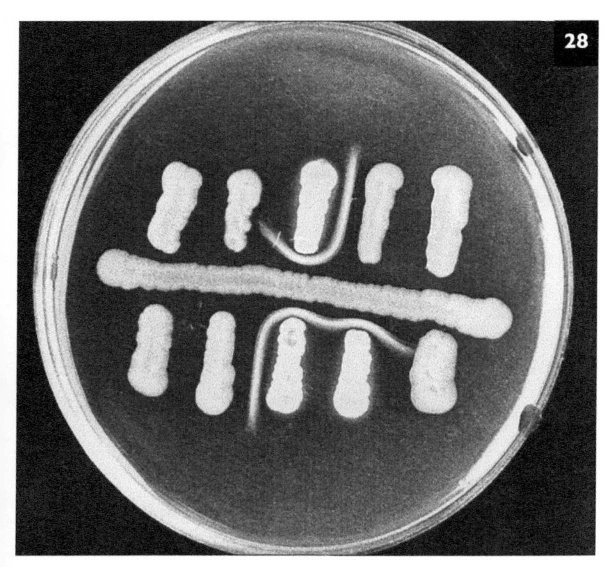

28 The white line test for the identification of *Ps. tolaasii.* The organism *Ps. reactans* is in the centre of the plate and the isolates to be tested are streaked either side (10 in this case). A white line of precipitation forms in the agar between isolates of *Ps. tolaasii* and *Ps. reactans*. There are three positive identifications on this plate. We are grateful to Dr W.C. Wong for the use of this figure.

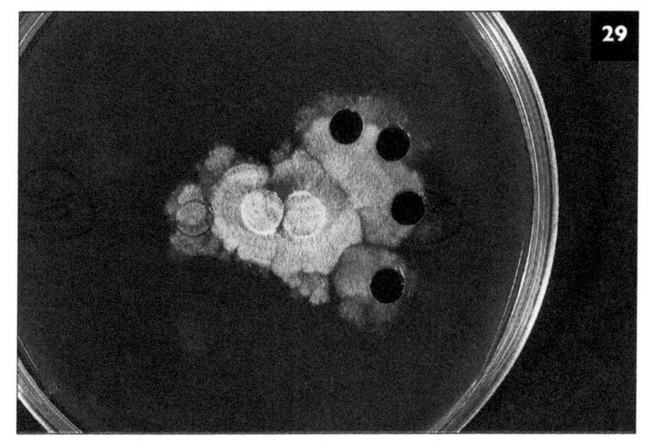

29 Transfer of mushroom virus by anastomosis. Virus-affected culture (round white block in centre left) has anastomosed with the healthy culture (white round block centre right), and the resultant growth on the right is tested from the leading edge (four agar plugs taken) and these show the same slow growth rate as the original virus culture if anastomosis has occurred and the virus been transferred.

the juice to be examined is placed on the grid, the virus particles are attracted to the antiserum. In this way, the virus particles are concentrated and are easier to find.

Techniques based upon the measurement of double-stranded ribonucleic acid (dsRNA) are now used, and these enable La France disease to be accurately identified. The techniques depend upon mushroom viruses being predominantly double stranded RNA. The virus in the affected mushrooms is extracted and the dsRNA separated from their protein coats. It can then be identified and quantified. A technique known as polyacrylamide gel electrophoresis (PAGE) has been successfully used for some years. The extracted dsRNA is diffused along a gel column with the aid of an electrical field, and according to its molecular size, reaches a point of equilibrium. By using a stain, fractions or bands of dsRNA can be seen in the gel. When La France virus is present, nine bands are recognized.

Other nucleic acid technologies have been used more recently in virus identification. A technique called reverse transcriptase polymerase chain reaction (RT-PCR) is capable of recognizing very small fragments of virus nucleic acid (dsRNA). This method has been used to check mycelial cultures, as well as mushrooms, for the presence of La France disease and is quicker, more specific and more sensitive than PAGE. It has particular value in phase III compost production as compost can be checked before it is distributed to customers. Like PAGE it requires specific equipment. The result of the test is either positive or negative, and there is no quantification of the amount of virus present.

Double-stranded RNA analysis, in particular a modified PAGE technique, has been used in the recent studies of Mushroom virus X disease. Up to 26 different dsRNA bands have been found where this disease syndrome occurs. These dsRNAs cannot be seen with the electron microscope because of the absence of a protein coat. Generally dsRNA is present in much smaller quantities than it is when La France disease is present. Analysis of Mushroom virus X disease is therefore dependent upon starting with about twice as many mushrooms as are used in analysis of La France disease. So far it is not clear which, if any, of the bands cause the disease symptoms, but if they do, they may represent a new virus group. Such a group, without coat protein, has been found in some green plants and also in some other fungi.

Viroids, Mycoplasmas, and Rickettsias

At present none of these groups is known to be associated with any of the mushroom diseases described in this book. However there are some diseases, such as Mummy disease, where the exact cause is still in doubt, and there have been suggestions that fastidious bacteria (a group of bacteria that are difficult to grow on artificial media) may be implicated. When investigating new diseases, all known groups of pathogens should be considered.

Viroids are identified using techniques similar to PAGE in which the ssRNA (single-stranded) is extracted and examined after gel electrophoresis. Viroids are small in comparison with viruses, so the amount of ssRNA extracted is much less. The ssRNA is generally of one type, so that one band occurs in the developed gel. Viroids differ from viruses in that they have no coat protein so cannot be seen using an electron microscope. In this respect they have similarities with the dsRNA associated with Mushroom virus X disease, except that viroids have one large band of RNA.

Mycoplasmas or phytoplasmas (those in green plants) are identified by specialized techniques including visual microscopy accompanied by specific staining. They are plasmodial-like organisms (without a fixed shape).

Rickettsias and fastidious bacteria are very difficult to culture, and their presence is usually detected by microscopy, specific stains, fluorescent antibodies, and DNA technology (PCR). Fastidious bacteria, which include some species of *Pseudomonas*, but not those known to be associated with mushroom culture, are so far, only found in the vascular tissues of plants. They are usually transmitted by vectors, such as sucking insects. Although bacteria are frequently found within fungal mycelium, they have not yet been identified as this type.

Abiotic disorders

There is no single method of identification. All possible factors must be considered (*see* pp. 27–29).

CHAPTER 3

Effective Pest and Disease Control

Introduction

It is important that modern commercial mushroom production is essentially pest- and disease-free. The means of achieving this has changed with production methods, and market requirements. In order to reach current high standards all available means of pest and disease control are used. The integration of these requires a sound knowledge of the biology of the pests and pathogens.

The most important single factor in an integrated programme is the effective management of pest and pathogen populations. When populations are out of control, major crop losses soon follow. Understanding the important factors in the management of populations, and the need to integrate them, is the subject of this chapter.

The components of an integrated programme of control are summarized in *Table 1*.

TABLE 1 **The components of an integrated pest and disease control programme**

The principles of pest and pathogen management

Entry
Containment
Elimination

Practices and operations

Identification and records
Filtration, ventilation and air movement
Disease removal
Harvesting hygiene
Disinfectant pads
Crop termination
Farm design
- Chemical control
  - application
  - degradation
  - pathogen resistance
- Power washing, disinfectants and disinfection
- Environmental control
  - composting
  - cropping
- Genetic control

Pest and pathogen management

The maintenance of healthy crops is the aim of integrated pest and pathogen control and the programme often referred to as 'farm hygiene', although in mushroom farming the term hygiene is generally used in a more restricted sense to refer to all means of control other than the use of insecticides and fungicides. The programme has three very important aims. First the exclusion of harmful organisms, second the containment of those that have not been excluded, and finally the elimination of any that remain. It requires a continuous effort on the part of management and lapses will quickly result in increases in pest and pathogen populations.

Exclusion

Exclusion implies an external source and although this is always possible, particularly for the initial introduction of pests and pathogens, the farm is generally the most common source. An effective programme will minimize the risk of introduction from outside the farm as well as reducing the chances of spread on the farm from affected to clean crops. On the farm, exclusion applies to all stages of production from the start of compost preparation to the end of cropping. Means of exclusion include absolute filtration, well sealed phase II rooms, dust and fly filters on cropping houses, effective disinfectant mats, and strict observance of all hygiene measures aimed at the prevention of spread by personnel. Very occasionally a specific source of pathogens or pests is identified, such as contaminated casing materials. This is a known primary source of introduction of *Mycogone perniciosa*, *Cladobotryum dendroides* and cecid flies. In addition, the now common practice of preparing compost on one site and using it elsewhere increases the chances of the introduction of pests and pathogens. Dramatic recent examples of diseases introduced in this way have been Trichoderma compost mould and Virus X disease. But it can also improve control in phase III compost, particularly of flies and sometimes moulds and virus diseases.

Containment

Containment is an ongoing operation on most mushroom farms. Its objective is to minimize spread, particularly from crop to crop. The main means of containment include strict hygiene at harvesting, the identification and correct treatment of disease patches, the use of protective clothing, the use of disinfectant pads, and air filtration. Fungicides and insecticides also play a role, as does the use of environmental controls used to slow down the development of diseases.

Elimination

Elimination of pests and pathogens from a farm following a severe attack is generally not completed in one operation but by using a number of processes over a period of time. The use of pesticides as well as some of the procedures used for containment are part of this process, but the most effective of all the processes is cook-out at crop termination. Generally heat cook-out achieves a total kill. Over a period of time, when the process is repeated for each growing room, the pest and pathogen populations on the farm are reduced to very low levels if not eliminated, particularly if containment is working well. An additional useful tool is the early termination of crops when pest or pathogen populations are high. This is often done after the second flush of mushrooms and is very important, particularly for organic producers who do not use pesticides.

Practices and operations

There are a number of practices and operations that are used in exclusion, containment, and elimination and together with others they are the basis of the management of pest and pathogen populations, forming the components of farm hygiene.

Pest and disease recording

Pest and disease records are an essential part of monitoring pest and disease incidence. The records provide vital information on the efficacy of the farm's hygiene programme, and often indicate whether or not there is a need for pesticide use. Identification is an important factor in the compilation of records. Familiarity with symptoms of diseases and the recognition of pests are generally sufficient, but occasionally laboratory examination is needed.

Simple techniques are used to measure the quantity of pests and diseases. Pests, in particular flies, are recorded by trapping them on strategically positioned yellow sticky traps. These are sheets of plastic approximately 15 by 30 cm which are covered by polybutene glue. They are placed in growing rooms, spawning areas, corridors, and other positions where recording flies will be helpful. Putting the traps near to a light source increases the numbers caught. The traps are changed for every crop in a cropping house. The numbers of flies caught are counted at regular intervals (at least weekly), and the rate of increase assessed. The flies can also be identified. In this way it is easy to see if the fly population is increasing, and the appropriate control measures can then be taken.

Disease teams can record the numbers and type when affected patches are covered, and these records are entered on the log sheet for the crop. The weekly totals give a clear indication of the numbers of disease 'pieces' in each crop, and of the disease levels on the farm.

Filtration

Filtration has become of increasing importance over the past decade. The removal of flies, fungal spores and mycelial fragments from the air used to cool compost or to ventilate the crop reduces the risk of pests and diseases developing.

All levels of filtration prevent access by flies as long as the mesh size does not exceed 0.3 mm. The removal of spores of pathogens, moulds or mushrooms is more difficult. In the past, it has been considered appropriate to filter air to 2 microns at 98% efficiency (**30**). This generally is known as absolute filtration. How many of the 'absolute' filters that have been fitted actually meet this specification is a matter of conjecture.

30 A large absolute-filter assembly with the filter units projecting. This assembly is in the roof space above the units requiring filtered air.

Filters are now rated by number into coarse 1–4, fine 5–9, and ultra fine 10–14, known as HEPA. Which of these is appropriate for a specific situation should be established by discussion with the manufacturers. For phase II or III compost production, absolute filtration is essential. Until recently, only phase II and phase III rooms were fitted with absolute filtration. Other rooms, if filtered at all, would have been equipped with coarse or dust filters. There are now good reasons to use absolute filtration on inlets and very importantly exhausts in other areas, such as spawning halls, central working areas and even cropping houses. Such filtration in these areas helps to limit the spread of virus diseases and some fungal pathogens.

31 A dust filter over an exhaust port of a growing room. The filter is black with dust and mushroom spores.

The specifications of filters and their effects on air flow require careful analysis if the protection required is to be achieved without reducing yields. In some instances, dust filters are adequate, even if the pathogen spores are small enough to appear to warrant finer filters (**31**). A good example of this is where the spores, such as those of *Verticillium,* are carried on larger dust particles. Mushroom mycelial fragments, important in the introduction of virus diseases, may also be removed by dust filters although they vary in size and some are small enough to pass through such filters. Each disease requires individual consideration when deciding upon a filtration policy. *Table 2* shows the measurements of the spores of the most common pathogens. The smallest spores are those of some of the *Penicillium* species.

Where filtration is critical, it is important that all the air used passes through the filters. Only too commonly, air is able to bypass the filters through holes in ducting, or because of poor fitting of the filters.

The use of such high specification filters can impair air flow due to the resistance exerted by the filter. This can adversely affect the crop.

TABLE 2 Spore measurements of the commonest pathogens of *Agaricus bisporus*

Pathogen/disease	Spore type	Spore size (microns)
Verticillium fungicola var. *fungicola* Dry bubble	Conidia	3.8–7.2 × 1.2–2.4
Verticillium fungicola var. *aleophilum*	Conidia	4.5–8.0 × 1.5–2.5
Verticillium psalliotae	Conidia	6.0–10.0 × 2.0–3.5
Mycogone perniciosa Wet bubble	Conidia Aleuriospore	13.6–17.5 × 3.7–6.1 18.3–23.7 × 19.9–26.1
Cladobotryum dendroides Cobweb	Conidia	22–27 × 7.5–9.0
Cladobotryum mycophilum	Conidia	15–32 × 7.5–12.0
Diehliomyces microsporus False truffle	Ascospores	5.0 × 7.0
Trichoderma aggressivum Trichoderma compost mould	Conidia	2.4–3.2 × 2.2–2.8
Trichoderma viride	Conidia	2.8–5.0 × 2.8–4.5
Trichoderma koningii Cap spot	Conidia	3.0–4.8 × 1.9–2.8
Penicillium chermesinum Smoky mould	Conidia	2.0–2.5 × 1.5–2.0
Agaricus bisporus Mushroom	Basidiospore	5.0 × 7.0

Ventilation and air movement

Ventilation and air movement and, with them, the evaporation of water, play a very important part in the control of diseases that cause spots, e.g. Bacterial blotch and Cobweb disease. It is essential that the surfaces of mushrooms dry quickly in order to prevent the development of both fungal and bacterial pathogens. The theory and mechanism of this important means of disease control is described in the section on cultural and environmental control (*see* p. 58).

In the past it has generally been thought desirable to over-pressure rooms. By using a positive pressure in an area the entry of unwanted spores can be prevented, but the danger of this system is that over-pressure can simply result in effective dissemination from the pressurized area to other parts of the farm. This is well illustrated with Cobweb disease where air-borne spores from an affected crop can be distributed by the over-pressurization of the growing room. However, this problem can be minimized, particularly during the vulnerable period of harvesting, by restricting the times when cropping house doors are open.

Similarly, the over-pressurization of bulk spawn-running tunnels may allow the escape of minute mushroom mycelial fragments, which could be a potent means of virus dissemination to other spawn-running composts within the system. The same problem could occur in over-pressured phase III emptying halls, and over-pressured growing rooms at crop termination, where the crop has not been killed by heat. A similar means of distribution of mushroom spores from cropping houses can also occur.

Over-pressurizing rooms remains a valuable technique that can be used to protect vulnerable phases of mushroom cultivation. However, the possible dangers should be taken into account when making decisions concerning the over-pressurization of any areas.

Dust management

Dust, consisting of inorganic and organic debris including mushroom and pathogen spores and mycelial fragments, is ever present on mushroom farms. Air movement is a means of dust dissemination and filtration is one way of preventing it from entering sensitive areas. With some systems of cropping however, in particular the shelf system, filling and casing is often done with the compost fully exposed to dust from around the farm. In such circumstances, and to a lesser extent with all systems of growing, dust management is critical. Concrete areas, around the entrances to the cropping houses on shelf-farms, in particular immediately before filling, must be power washed and disinfected before cultural operation begins. For this operation to be effective it is essential that the concrete is in good condition.

Wherever possible, the compost or casing should be covered on the elevator. As soon as the compost is in its final position it should be covered with paper (**32**). Disinfectant mats should be positioned at the door to prevent contamination of the cropping house floor.

Disease removal

The successful treatment of patches of disease is critical for the control of some diseases. On all but the very smallest farms, this is done by 'disease teams'. Team members are commonly associated with and drawn from harvesting teams. The function of any such team is to recognize and isolate or remove any disease before spread occurs.

'Diseasing' is an important and skilled job and should be accorded both the respect and the authority the function deserves. Members of the team should be trained in the recognition of diseases or any disorders within the crop. The onus is on the recognition of the abnormal rather than on precise identification, but good disease teams will readily recognize and identify the common diseases. They should be provided with adequate means of illumination and must be aware that unless they rigorously practise hygienic

32 Spawn-running in compost protected with a covering of polythene. Paper is also used for this purpose.

principles they are potential spreaders of the diseases they are there to control (**33**). For this reason, and because disease teams often enter more houses than pickers, they should be supplied with protective clothing (overalls, gloves, aprons, sleeves), which is changed for every crop. On occasions, it may even be necessary for them to change gloves more than once within a crop. All team members must tread in a disinfectant foot-bath between every crop in order to avoid spore transfer on the soles of boots.

Unrecognized disease awaits dissemination by spores, flies, mites, water, or personnel. It is the sole responsibility of the disease teams to remove or isolate pieces of disease as soon as possible, and particularly before watering or harvesting begins. Generally, cobweb, *Verticillium* and *Mycogone* are the diseases most commonly treated by the disease team. The team works from the cleanest to the most severely affected crops. Most commonly the operation begins at the end of the first flush of mushrooms, but sometimes it is necessary to start disease removal before any mushrooms are ready to pick. The very early occurrence of disease indicates a lapse in hygiene which must be identified and corrected.

33 A powerful hand held torch used by a member of a disease removal team.

To ensure that disease teams operate efficiently the following points are essential:

- Identification: disease teams must be trained to recognize abnormalities and in particular the early stages of disease development.
- Illumination: must be adequate in order for disease teams to see the early stages.
- Disinfectant: disease teams must use a disinfectant foot-bath between crops.
- Isolation: disease teams must be trained in the use of appropriate isolation or removal techniques (*see below*).
- Authority: their work must be given priority and they must have the time and authority to complete it. Management must recognize the importance of the work of the disease team.
- Access: the disease team should have priority to work in a crop before watering or harvesting begins.
- Timing: as part of the disease control hygiene strategy, judgement is required on when inspection should begin and at what frequency. For instance, inspection before the first pick is not usually necessary, but occasionally is essential if disease has been found. Teams should not be used where success is unlikely, for example, when searching for Cobweb disease with a flush of open mushrooms that are completely covering a bed.
- Inspection: the frequency of inspection is normally once a day, but in some instances, for example with rapidly developing Cobweb disease, twice a day may be needed.

Areas of disease should be very carefully covered with salt in order to prevent further spread. How this is done is of major importance. Disease specimens should not be touched or handled, even with gloved hands. If accidentally touched, gloves must be changed immediately. Even when removed with great care, the dry spores of *Cladobotryum* are readily air-borne and *Verticillium* spores stick to everything that touches them. The following points are very important when treating disease on beds:

- *Verticillium*- or *Mycogone*-affected mushrooms should be carefully and completely covered with salt. The cover should be extended to an area about twice the radius of patches of the visible disease (**34a**). Free-flowing salt is usually considered easier to use than the large crystal type.

- All areas of Cobweb disease should be very carefully covered with damp paper towelling, ensuring minimal disturbance of the surface. The towel is then covered with salt starting from the edge of the colony and working towards the centre in order to weigh down the paper and trap spores (**34 b, c, d**).
- Large mushrooms with Mycogone and Verticillium (never Cobweb) are best physically removed because they are difficult to cover with salt.
- Physical removal requires very particular attention, due to its potential for disease spread. The disease should never be handled directly but should be picked-off with disposable paper towelling or disposable gloves, used once only.
- The removed mushrooms should be either bagged or, if the quantity is small enough, placed in a bucket with disinfectant. The area from which the disease has been removed should then be salted. Disposal of bags of diseased mushrooms should be done with great care.

Harvesting hygiene

Harvesting hygiene applies predominantly to pickers, but also to the disease teams and anyone else who may have intimate contact with crops.

Hygiene strategies for pickers are based on the assumption that, however efficient disease teams have been in their attempts to isolate or remove disease before the entry of pickers, some disease will be missed. By far the most spreadable disease is Verticillium, because of its sticky spores. However, Cobweb can also be spread by pickers, not only by disturbing spores, but also by transferring small pieces of Cobweb mycelium. In order to prevent inadvertent spread, equipment must be disinfected at regular intervals. This is often done between crops when moving from one crop to the next. Strategically placed dip tanks of disinfectant must be large enough to allow pickers to immerse completely all their picking equipment. Although the immersion time may be very short, the treatment will at least reduce the contamination if not eliminate it. It is very important

34 Covering cobweb disease with salt. (**a**) Carefully covering the disease patch with a wet paper towel; (**b**) sprinkling salt around the edges of the towel; (**c**) covering the whole area. Extensive patches of Verticillium (**d**) can be covered without using paper.

to choose a safe disinfectant for this purpose, and those containing quaternary ammonium compounds are often preferred. Buckets, knives, stands etc must be dipped before the new crop is entered (35).

Returnable plastic containers are frequently used during harvesting. These are made of durable plastic and they are cycled from the farm to the retailer and from the retailer to the farm. They are used by supermarkets for mushrooms and many other types of produce. Because they are recycled they represent a disease risk as they could be vehicles of transfer of pathogens and mushroom spores from one mushroom farm to another. The greatest risk of this occurring is when they are taken into the crop, and mushrooms are packed directly into them. These large containers may then be used to display the product on the shelves of the retailers before being returned to another mushroom farm. Similar containers are used to take wrapped pre-packs to market, though the risk of pathogen transfer with these is far less. Some retailers use tray washing lines to clean returnable containers. Large tunnel washers with a very large throughput process the containers in a very short time, sometimes as little as 30 seconds. The line may consist of three separate tanks: an initial wash tank usually containing hypochlorite solution at ambient temperature where loose debris and large particles are removed, followed by a caustic soda wash at 60°C or above, and finally a quaternary ammonium rinse at about 87°C. Although this procedure has been designed to eliminate human bacterial pathogens, it is likely that it will at least reduce the risk of mushroom pathogens and mushroom spores being transferred to another farm.

The key elements to harvesting hygiene are:

- Picking teams should work from clean to dirty crops, which is usually from the youngest to the oldest.
- Disposable gloves are used and should be replaced between every crop.
- Overalls should be laundered every day.
- Everything moveable in the cropping house which is touched by the pickers, such as knives, stools, picking racks, steps, and crates should not be taken to other houses unless these items have first been thoroughly cleaned and disinfected.
- All surfaces touched by pickers, such as radios and door handles including those in canteens, rest rooms, and lavatories, can become contaminated. These should all be disinfected.
- Returnable plastic containers should not be taken into crops unless they have been adequately cleaned.

35a

35b

35 (**a**) Harvester about to disinfect his harvesting equipment; (**b**) disinfection of harvesters' knives; notice the open basket placed inside the bucket of disinfectant which allows the whole of the knives, including the handles, to be immersed.

Foot-dips

Pathogens can undoubtedly be spread on footwear and foot-dips are used to prevent this. Foot-dips also remind personnel entering or leaving cropping houses of the need for hygiene.

Ideally disinfectant trays should be deep enough to thoroughly treat the soles of footwear. The disinfectant trays must never be allowed to dry out and must be routinely examined and topped up. Disinfectant trays can be avoided by personnel and not regularly used. Where this is possible it may give the impression that hygiene is not really important; it will also allow pathogens to leave contaminated areas and pass into vulnerable areas (**36**). Trays should be large and in such a position that it is almost impossible to not use them when entering or leaving a vulnerable area.

Extensive mats are sometimes used instead of foot-dip trays, and these, if large enough, are almost impossible to avoid. They have the additional advantage in that they can be used to disinfect wheels.

Foot-dips are not only for use by pickers and management but, very importantly, for disease teams as they move from crop to crop.

Disinfection of vehicle wheels

Not only feet but also wheels can be a means of introducing pathogens and pests into cropping houses and other sensitive areas. One place where wheels could be particularly important is where road vehicles are entering clean areas to load phase II or phase III compost. A trough sunk into the concrete at the entrance to the clean area must be deep enough for the tyres to be covered, and long enough to allow one complete revolution of the wheels. It must be regularly maintained with fresh disinfectant (**37**).

Fork-lift trucks bringing trays of compost into clean cropping houses can introduce debris on their wheels. An absorbent cloth soaked in disinfectant and long enough to allow one revolution of the wheels, can help minimize contamination of the floor when placed across the entrance to the house.

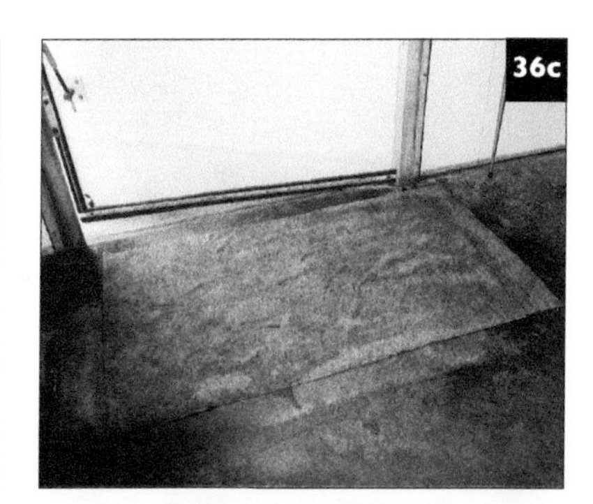

36 Disinfection pads at doorways. (**a**) Tray and pad in a prominent position but could be avoided because of its size; (**b**) a much larger tray and pad filling a doorway, but probably ineffective at times because of the damage to the tray which would allow the disinfectant to drain out; (**c**) disinfectant mat completely covering a doorway and very difficult to avoid.

37 A trough of disinfectant at the entrance of a compost production unit. Note that the trough is deep enough and long enough to completely disinfect the tyres of vehicles entering the unit.

Crop termination

Pest and disease levels almost invariably increase with time, reaching a maximum by the end of the crop. Effective crop termination is essential and is one of the most significant ways of reducing the overall population. It provides an effective break for the growing room, giving a clean start for the next crop. Populations of pests and pathogens in the last week of cropping are usually at a maximum and often represent the bulk of the population present on the farm at that time. In order to maximize the effectiveness of the cook-out, the extant crop should be treated by the disease team on the day before cook-out, with either the removal or cover of remaining disease. Salt cover also minimizes possible spore spread from the room when the steam is first introduced and air is forced out.

By far the most effective crop termination is heat treatment *in situ*, but this is not possible on many farms as growing rooms may not have steam available, or they may not be able to withstand regular treatment without severe structural damage. A common alternative is a specially constructed room, used exclusively for cook-out (**38**). Shelf farms must be able to be cooked-out *in situ*. When the crop is grown in bags, these can be moved to a cook-out room, but block growing presents problems as the crop is not easily moved from the shelves or racks used for cropping.

Cook-out in the growing room, not only heats the crop to a lethal temperature, it also 'cleans' all the surfaces of the building however inaccessible. The surface of the concrete floor is also effectively treated although the heat is unlikely to penetrate very far into cracks or joints.

38 Trays stacked in a cook-out tunnel ready for treatment.

Growers have found that a temperature treatment of 65–70°C for 9–12 hours is an effective treatment. This is well above the thermal death point reported for the organisms that need to be killed. For instance, most of the fungi are killed at 45°C with a treatment time of 10 minutes, while insects, mites, and nematodes, are killed at lower temperatures. A number of factors account for the discrepancy between the theoretical thermal death-points and the temperature used in practice. The most obvious is the speed of heat penetration and uniformity. In order to reach a lethal temperature in all the compost, it is necessary to treat some of it at a much higher temperature. This is particularly so if the compost is wet and dense. There is also the need for the heat to penetrate into tray timbers as some pests and pathogens may be within the wood or in joints and cracks not only in the wood, but in the concrete floor. Mushroom spores can also be washed into such places.

Another factor is the greater resistance of dry spores to heat treatment. As spores dry, they become more resistant to heat, and when totally dry they may be extremely difficult to kill. This is generally not a problem at cook-out as the crop is not totally dry. However, differences in the thermal death-point temperatures that have been reported could be accounted for by the water content of the spores at the time the tests were done. This is particularly relevant for mushroom spores where considerable variation in thermal death-point temperatures have been published. Taking all these factors into account, a temperature of 65°C should be adequate to kill all pests and pathogens, as well as mushroom spores. The length of time of the treatment is governed by the circumstances of the farm. On a modern farm, it should not be necessary to maintain this temperature for more than 9 hours. On older farms a 12-hour period at 70°C may be needed as heating is not uniform and heat loss is greater.

Where there has been a persistent disease problem such as dry bubble, the use of formaldehyde together with heat treatment can further increase the efficacy of the cook-out, although this should not be necessary if the cook-out is done properly (**39**). The formaldehyde is used as formalin at 4 ml/m^3 (*see* p. 56).

In extreme cases, in which pest or disease levels are very high at the end of a crop, it can be extremely beneficial to advance crop termination to, for example, the end of the second flush, rather than at the end of the third. This can reduce pathogen populations by over 90%.

If disease has got out of control during the last week, great care must be taken if the crop is to be moved from the growing room to a cook-out room. Treatment with a disinfectant is an essential before removal of all crops not cooked-out *in situ*, making certain that every part of the surface is disinfected. During and after disinfecting, the cropping house should be sealed for at least 1 hour to allow air-borne spores to settle. After this time the crop can be removed to the cook-out room.

Important points for steam cook-out are:

- Steam cook-out is the most effective method especially when done *in situ*.
- Treat all disease up until the day before cook-out.
- Maintain a temperature of 65–70°C for 9–12 hours depending on circumstances.
- Where the crop is to be moved to a cook-out room, shut off the house fans, spray the surface of the crop with a disinfectant, and leave the house closed for at least an hour after treatment to allow air-borne spores to resettle.
- Check cook-out temperatures using probes in the compost. Thermometers or heat-sensitive paper can be placed in various parts of the room, particularly those parts where lethal temperatures are most difficult to achieve.
- Where cook-out temperatures are not reached, the crop should either be reheated or treated with chemicals.
- After removal of the crop, power wash with water to remove crop debris. Avoid contamination of other areas by the run-off and splash from this washing.
- No further cleaning should be necessary where cook-out was *in situ* although disinfection of the concrete floor, particularly if there are many cracks and joints, may be advisable.

Where steam is not available and the crop is not terminated *in situ*, it must first be sprayed with a disinfectant and removed from the house before being taken well away from the farm. On no account should a crop sprayed with disinfectant be left for more than 12 hours, as mushrooms and disease will continue to develop because the treatment is to the surface only. If it is sprayed but cannot be emptied on the same day, the spray treatment must be repeated the day that it is emptied. Generally the disinfectant causes the mushrooms to discolour and the likely efficacy of the treatment can be gauged by the area of brown discolouration on mushrooms left on the bed (**40**).

39 Equipment used for formalin fumigation. The large metal container holds the formalin and a bottle gas supply provides the fuel for the heat for vaporization.

40 Mushrooms after disinfectant spray-off at crop termination. Notice the uneven distribution of the brown discolouration indicating uneven application of the disinfectant. This poor application would result in some diseased areas not being treated effectively.

It may be necessary to repeat the treatment on some poorly covered areas. Following spray treatment (often referred to as spray-off) the crop is carefully removed and the floor power washed with water. After washing, the house must be thoroughly dried before being fumigated with Formalin (*see* p. 56). It is then kept closed for 24 hours before airing and finally washing the floor with a disinfectant.

The main points of 'chemical' cook-out are:

- Spray the remaining crop with a disinfectant not more than 24 hours before emptying.
- Carefully remove the crop to avoid the deposition of debris.
- Power wash the floor.
- Dry the cropping room after power washing,
- Fumigate with formalin, leaving the room closed for 24 hours at a temperature of at least 15°C.
- After fumigation disinfect the floor.
- Keep the doors closed after cleaning has been completed until the next crop is brought in.

Emptying

The removal of used compost from the growing room should present no pest or pathogen risk to other crops on the farm if steam cook-out has been totally effective. This, unfortunately, is not always the case. Also, where termination has been done solely with the use of disinfectants, pests will largely have survived, and mushroom mycelium within the compost being unaffected will constitute a major risk. A knock-down insecticide used the day before emptying will eliminate most adult flies. Disposal of the used compost should always be treated as a high-risk procedure. Fragments of mycelium and mushroom spores, which are potential sources of virus diseases, as well as pathogen and mould spores, may have survived the disinfection (**41**).

After emptying, it is important to clean the area by thorough disinfection. It is also important to prevent, or at least minimize, the entry of debris into crops by making certain that nearby houses are closed and air intake is minimized during emptying, particularly if the emptying process is taking place near to crops. An additional hazard is the accumulation of compost debris in areas (such as between cropping houses) not washed down at the end of the process. Even if this debris is free from pests and pathogens, it can provide an ideal breeding area for flies, in particular Sciarids. Used compost should always be taken well away from the farm.

41 Used compost being loaded to be taken off the farm. This operation can result in pest and pathogen spread if the compost is not effectively cooked-out.

Empty trays

A temperature treatment of 60°C will clean all surfaces of empty trays. This treatment should not be needed if cook-out was fully effective, but is sometimes used if the cook-out temperatures were not fully achieved or the farm has a persistent disease problem. Where the crop is not cooked-out, the trays should be chemically treated. The efficacy of tray treatment is considerably aided by lining the trays with polythene (**42**). When the polythene is removed at emptying, the trays are relatively free from compost and debris. They can then be dipped in a disinfectant (only those registered by the appropriate authority for this purpose).

The trays must be totally submerged and all surfaces thoroughly wetted (for details of suitable disinfectants *see* pp. 54 and 56). Disinfectants clean the surfaces only and do not penetrate into the wood. In general, the longer the wood is exposed to the disinfectant the more effective it is likely to be. A treatment time of 10 minutes should be considered

42 Polythene lining of the trays overlapping the edges. Contamination of the wood is less likely to occur when the trays are lined and the trays are also less contaminated with compost when they are emptied.

43 Clean trays being stored before re-use. It is important to position them away from the air exhausts from cropping houses and preferably on the windward side of the farm so that dust from the farm is not blown onto them. When large numbers are involved the ideal may be difficult to achieve.

to be a minimum although this will, of necessity, be made up of a thorough dip followed by a period when the trays are still totally wet. Quick drying of disinfected trays should be avoided.

The correct concentration of the disinfectant must be used, and must be regularly checked, because the presence of copious amounts of organic matter may reduce the efficacy of the chemical. Often, the large tank used for dipping has additional disinfectant added each week in order to maintain an effective concentration. If this is done with no check on concentrations, it can result in either excessive use of product, which may lead to toxicity to the mushroom mycelium in contact with it, or conversely an ineffective disinfection. Phenolic disinfectants are a common choice and it is not easy for growers to check their concentration. Disinfectant lost during the dipping process must be replaced with more disinfectant at the appropriate concentration. The supplier of the disinfectant should be consulted for information on testing the concentration. The concentration of triazole fungicides included in the Safetray products can be checked for concentration in the tank using a refractometer.

At intervals, depending upon the amount of debris in the tank, the solution must be replaced. Safe disposal of the used solution can present problems and advice must be sought on the best way to do this. Phenols can cause considerable environmental damage, even in very low concentrations.

Once trays have been treated, either by heat or by chemicals, they must be stored in an area where they will not become contaminated by air-borne spores or fragments of mycelium (**43**). Spores of *Agaricus bisporus* are the most likely contaminants, although those of some pathogens may also be a risk. Ideally, treated trays should be stored well away from cropping houses, and preferably on the windward side of the farm.

Important points for tray cleaning:

- Heat treat empty trays at 60°C for 6 hours. The addition of formalin at 4 ml/m^3 (*see* p. 56) will increase the efficacy of the process.
- If possible, line trays with loose plastic before filling to minimize the carry-over of compost.
- When disinfecting trays, make sure they are thoroughly wetted and remain wet for at least 10 minutes.
- Once trays are clean, store them in an area where they will not become contaminated by air-borne spores.
- Check the concentration of disinfectant in the dip tanks weekly.
- Discuss the safe disposal of disinfectant with the local authority.

Farm design

Mushroom farms often change over a period of time and may be far from ideal when factors affecting pest and disease control are considered, but new farms can be designed with effective control in mind. The principal aim of the design is to separate cropping areas from preparation areas both physically and also in terms of traffic flow. Compost must move on a pathway from phase I to phase II, to spawning, to spawn-running, to cropping in such a way that the risk of contamination by pathogens, pests, or mushroom mycelium and spores is minimized. Such an ideal is not always possible, even when starting with a new farm, as the topography of the site may dictate where some of the buildings are situated. However, the principle of keeping clean away from dirty should be followed if at all possible.

Chemical control

The chemicals used by mushroom growers as part of the integrated control strategy of pests and pathogens include fungicides, insecticides, acaricides (collectively referred to as pesticides), and disinfectants.

There are very few pesticides registered for use on the mushroom crop in any country, and those that are, are used very widely (*see* Appendix 1).

To obtain registration for the use of a pesticide, its safe use and efficacy must be proven to the satisfaction of the registration authorities. Although disinfectants are not currently subjected to the same scrutiny they must not be used on crops, or containers, such as trays, that are in direct contact with crops unless specifically registered for such use. Their use must always comply with safety regulations. Increasing emphasis is placed on safety in the environment as well as for the operator and mushroom consumer.

Pesticides must have a wide degree of differential toxicity. This is a particular problem with fungicides, as many potential materials could adversely affect mushroom crops. If a crop is overdosed, even with a registered material, it is likely to result in mycotoxicity which, in mild forms, may delay cropping by a day or two, or if very severe, can prevent cropping altogether.

Application of pesticides

It is vitally important to adhere precisely to the instructions on labels on pesticide containers. The correct concentration, volume, and timing of the application must be precise if the best performance is to be obtained from a product. Similarly the operator safety instructions must be very closely followed, as many products used in pest and disease control are dangerous especially when undiluted. The safe disposal of excess product and empty containers is essential in order to prevent environmental pollution.

The most common way of applying pesticides to mushroom crops is by watering them onto the casing surface (up to 2 litres per m^2) in place of, or as part of, routine watering. 'Watering-trees' are in common use and these are wheeled between the rows of shelves or trays and the pesticide is delivered onto the casing surface (*see* **12**). Commonly a pesticide is applied on the day of casing or up to 7 days after, depending upon the product. Sometimes further applications are needed between flushes of mushrooms. The use of watering-trees will often result in pesticide going to waste and due allowance must be made when calculating the rate of application. Watering-trees are also quite frequently uneven in their delivery pattern. It is important to check at frequent intervals both the volume delivered and the delivery pattern.

A common alternative to this system is the use of a hose pipe fitted with a very fine-mesh rose delivering from a spray tank (**44**). This enables the operator to cover all surfaces and, with tray-crops, to spray around the legs of the trays. It is also an effective way of applying pesticides to bag crops.

Tray-crops can be sprayed on the casing line at the time of casing. A boom is fitted onto the line which delivers pesticide onto the casing surface as the tray passes. The advantage of this system is the even cover of the casing surface, which is much more difficult to achieve when the trays have been stacked in cropping houses.

In some countries insecticides can be applied to the compost for the control of flies. They are usually applied when compost is spawned, but it is important to avoid direct contact between spawn and insecticide.

Some fungicides can be mixed into casing before the casing is applied to the crop. The advantage of this method is that, if well mixed, it gives even distribution of the product. The disadvantage is that with some of the heavier wetter casing materials, it is very difficult to mix the fungicides adequately. Also, for the control of diseases such as wet and dry bubble, some product is wasted when it is mixed throughout the casing, as it is needed in the top half of the casing only.

Fly-control may be applied using a pesticide fog, or very fine mist, or by the use of pesticide smokes. With these methods of application the correct calculation of the volume of the room to be treated is essential.

Disinfectants are usually applied at high volumes or, in the case of formalin, by fogging. To do this safely, the operator must wear full protective

44 (a),(b) Application of water and/or pesticide using a hose pipe and a fine rose.

clothing, and use the appropriate filter for the respirator. Thermal fogging with formalin has the advantage that it can be done from outside the room to be treated, and the chemical is more effective in the vapour phase than when it is applied as a dilute solution. In order to get the best results from thermal fogging, the surfaces of the treated room must be dry and there must be no free water on the floor. A minimum temperature of 10°C is required, although a higher temperature of at least 15°C is likely to give better results. In the UK the maximum permissible amount of formalin that can be fogged is 4 ml/m^3. This rate may not apply in all countries and it is important to check the local rate of usage.

Concentration of pesticide in practice

The actual dose of pesticide applied depends upon the evenness of application and the accuracy of delivery. Investigations into the causes of apparent failures of pesticide to produce the expected level of control have demonstrated losses of product of up to 40% (of the calculated amount). It was concluded that this loss was due to a combination of factors such as (i) spray applied to floors and tray timbers; (ii) pesticide left over at the end of spraying; and (iii) material that had settled in the spray tank due to insufficient agitation. In other situations, unevenness of application can result in areas of the casing receiving a 20–30% overdose or a similar underdose. Loss of product through failure to reach the target, combined with the inherent difficulties of achieving an even application of pesticides, can result in poor pest and pathogen control.

Degradation of pesticides

Pesticides break down after application until there is no longer any active ingredient detectable. The time taken for this to occur varies with the pesticide. Apart from acceptable residue levels in crops, products are only approved for use if their rate of breakdown is fast enough to prevent their gradual accumulation in the environment. To be effective in mushroom crops, an active ingredient must persist long enough to provide protection at the critical stages of crop development, while leaving very low level residues, if any, in or on the mushrooms.

Occasionally, there have been instances of unexpected control failures, which have been attributable to the rapid disappearance of the active ingredient from casing. This is known as enhanced degradation and is a particular problem with some products. In mushrooms, the carbendazim/benomyl fungicides are liable to show this, particularly after repeated use. Bacteria in the casing appear to play a role in this process. Thiabendazole, a fungicide with a similar mode of action to that of the carbendazim fungicides, is much more resistant to enhanced degradation, but seems to be more likely to become ineffective because of the development of fungicide resistance. Recent work has shown that prochloraz manganese can be degraded in the spray tank by bacteria. In surveys it was found that this was not an uncommon phenomenon and substantial degradation could occur before the fungicide was applied. In such situations the spray tank must be thoroughly cleaned and steralized to remove the bacterial population before using prochloraz manganese.

Pesticide resistance

Pests and pathogens can become less sensitive to the chemicals used to control them and may also become resistant. Resistance or decreased sensitivity results in loss of control. Pesticide resistance is a common problem which often follows the indiscriminate and widespread use of a product. The selection pressure on the pest or pathogen population is then considerable.

Sciarid fly resistance to the organophosphorous insecticides is now widespread. Resistance of sciarid flies to diflubenzuron has also been recorded in the USA and UK. In the laboratory, resistance is determined by feeding studies and expressed as LD_{50} or LD_{95} values. These are 50% or 95% of the concentration which is lethal to all of the flies tested. Once resistance strains have evolved, the product becomes totally ineffective.

Fungicide resistance is usually expressed as an ED_{50} value (effective dose). This is the concentration of the fungicide that reduces the growth of fungal mycelium of the pathogen by 50% in tests on agar (*see* **27**). Sometimes an ED_{95} value is quoted, and this is very near to the total toxicity or lethal point for the fungus. To be useful on mushrooms, a fungicide must be considerably more toxic to the fungal pathogens than to the mushroom. A good example of a wide differential toxicity was the fungicide Benlate (active ingredient benomyl), where the ED_{50} value for the pathogen *Verticillium fungicola* was less than 1 ppm and for the mushroom 50 ppm. When resistance developed, the ED_{50} of the pathogen changed to 100 ppm. The complete failure to control this pathogen followed a short period of intensive use of Benlate.

Within *Cladobotryum* species and strains, there is a complex of resistance, partial resistance, and sensitivity to fungicides currently available. Even within the benzimidazoles there is variation in the sensitivity of *Cladobotryum*, with many more isolates being resistant to thiabendazole than to benomyl. In 2005, survey reports from the USA demonstrated the reverse situation with *Trichoderma aggressivum* var. *aggressivum*. Isolates were found that were resistant to benomyl and thiophanate methyl, but sensitive to thiabendazole.

Prochloraz manganese (Sporgon and Octave) is now widely used for control of Dry bubble. Its differential toxicity is not as wide as that of benomyl. Some isolates of *Verticillium fungicola* have been found that are not totally resistant but are less sensitive to this fungicide. Although they are still controlled by the application of prochloraz manganese, the duration of effective control is reduced. Increasing the concentration by decreasing the amount of water used during application can help to improve the control of these less sensitive strains but there is an increased danger of mycotoxicity.

Strategies to prevent the development of resistance depend upon reducing the exposure of the pest or pathogen to a particular active ingredient. In this way the selection pressure is reduced. Fungicides and insecticides should not be used as a routine but restricted to those times when their use is most necessary. In no circumstances should fungicides be included in additives of any type such as supplements, as the widespread and general use of these will create a continuous selection pressure which, in the end, will almost certainly result in fungicide resistance to the active ingredient.

The selection pressure on any one active ingredient can be reduced by using a mixture of more than one active chemical. The mushroom industry has generally not had the luxury of this choice so, although laudable, this means of resistance management is not always possible. Mixtures of two products from the same active ingredient group, for example the benzimidazole fungicides, are not beneficial as they behave as one product.

To ensure that insecticides and fungicides are as effective as possible, consideration should be given to the following factors:

- Evenness of application, which can be checked by accurately measuring the spray distribution and the area or volume to be treated.
- Resistance/sensitivity status of the isolate of the population of pests or pathogens, which can be established by laboratory tests. This is particularly relevant for *Cladobotryum* and sciarid flies before selecting the appropriate chemical.
- The indiscriminate use of insecticides or fungicides which will increase the selection pressure on the pest and pathogen populations and will favour the development of resistant forms.
- The routine use of supplements or other materials that contain pesticides, which will increase the selection pressure and thereby the chance of resistance developing to the pesticide used.

A combination of factors, such as difficulties in even and adequate dosing, the appearance of pathogen strains with reduced sensitivity, partial or outright resistance, in addition to various degrees of enhanced degradation of the pesticide, make it unlikely that chemicals alone will always achieve effective control.

Power washing

Power washing is often the first stage of cleaning a surface and is frequently done before disinfecting. It removes large amounts of debris and spores from a washed area. The smoother the surface the more effective power washing is likely to be. By power washing, the bulk of the organic material is removed and this will enable more effective treatment with disinfectants. But power washing, like spraying, causes small particles including spores to become airborne, and a settling time may need to be left after the washing operation.

It is important to eliminate cracks and crevices in concrete as these generally get filled with debris and possibly spores of pathogens. If debris is power washed out of the cracks, it can be splashed onto nearby surfaces. This is particularly relevant when such cracks occur in cropping houses and debris is splashed onto the surface of the mushroom beds. It is often the lower beds that are affected in this way, and it is not uncommon to have a disease gradient with most disease on the lowest beds, and less with distance above the floor.

It is common practice to power wash a house after casing in order to remove excessive casing which has fallen onto the floor. This will splash casing fragments and with it pathogen spores from the floor onto the bed surfaces, so must be avoided or only done on floors without cracks.

The water used in power washing is often drained from the site into a holding tank where other excess water is also held, for instance from the compost yard. This water often contains large amounts of organic material and unless it is well aerated it very quickly becomes anaerobic, especially in hot weather. It is very important to aerate such water, often referred to as 'goody water', in order to minimize smells on the farm. If the waste water is to be used during the initial stages of composting it should be analysed at intervals for its nutrient content, and also for the presence of the disinfectants used on the farm. If phenols are used to wash down roadways etc, the concentration in the waste water can rise to a level where it can interfere with successful composting. The soluble salt level in the waste water, if high, can also be a problem.

Disinfectants

A disinfectant is an agent, generally with a very broad spectrum of activity, which destroys or inhibits organisms. They are particularly effective for the elimination of fungal spores and fragments of mycelium but are not as effective against pests. In the mushroom industry, disinfectants are manufactured chemicals although there is no reason why they should not be natural products. They are as likely to be as toxic to the mushroom crop as to the organisms that cause diseases. In this respect, they differ from the approved pesticides, which are much more toxic to pests and pathogens than to the crop. This lack of differential toxicity is a very good reason for not applying disinfectants to crops, but it is not the only one. Disinfectants are not always covered by the pesticide legislation. But they must always be safe for the operator to use. They cannot be used on the crop directly or on any part of the crop unless they conform to the pesticide legislation. For this reason the treatment of casing, compost, trays, and crops with a disinfectant is not permissible unless it has been subjected to the same rigorous investigation and tests as a pesticide.

Removal of as much debris and organic material as possible from the area to be disinfected is very important if the best results are to be achieved. In this respect power washing must be considered to be a vital part of the process.

The efficacy of a disinfectant varies with the product, the concentration used, and the exposure time. Few if any disinfectants used in the mushroom industry at the recommended concentrations will kill spores if the treatment time is only a few seconds. More often a 10-minute treatment is required, and in mushroom farming such a treatment time is difficult to achieve if not impossible for many operations. Disinfection may not therefore give a total kill, but will reduce the population of pathogens. Continuously repeating the processes of disinfection on farms is a way of minimizing the risk of large populations occurring and hence severe disease.

The ideal disinfectant

The ideal disinfectant for a mushroom farm should have most if not all of the following characteristics:

- A wide spectrum of biological activity covering the whole range of pathogens that may be present.
- Rapid action time to allow effective treatment even when the exposure period is short.
- Effectiveness in the presence of organic material, which is an essential on a mushroom farm where compost and casing contamination of surfaces is commonplace.
- Effective at alkaline pH, because of the use of calcium carbonate in mushroom culture.
- Must have no smell or a pleasant smell, and one that is not transferred to mushrooms.
- Must be stable and have a long storage life under normal farm storage conditions.

- Must be operator friendly and convenient to use.
- Must be environmentally friendly with no toxic breakdown products and also easy to dispose of.
- Must be marketed in easily and safely disposable containers.
- Must be economic to use.

The chance of finding a disinfectant with all these properties is not very likely. The choice of product is therefore a compromise. Products vary in their biological strengths and weaknesses, and by choosing a combination of disinfectants, it is generally possible to do an effective job.

In some circumstances narrow-spectrum materials can be useful where there is a specific problem. For example, an acaracide for the control of mites can be used as a disinfectant when mites have become endemic on farms.

Types of disinfectant

There is a wide range of chemicals used as disinfectants. They are broadly classifiable into chemical types (*Table 3*). Before using any product,

TABLE 3 Major groups of disinfectants and their general properties for the control of mushroom pathogens

Chemical group	Bacteria	Fungi	Factors influencing efficiency
Alcohols	S	S	pH independent
Ethyl			OM independent
Isopropyl			Skin irritant
Aldehydes	S	S	Skin irritant
Formaldehyde			Possible carcinogen
Glutaraldehyde			OM independent
Chlorine	S	M	pH dependent
Sodium and calcium hypochlorite			OM inactivated Light inactivated
Chlorine dioxide			
Iodophors	S	S	pH dependent
Organic iodines			OM inactivation slow
Phenols	S	S	Best at neutral pH
Substituted phenols including alkylphenols, cresols, xylenols			OM independent Non-ionic soaps inactivate
Quaternary ammonium	M	S	pH best above 7
compounds (QACs)			OM inactivated
			Anionic soaps
			Inactivate
Surfactants	S	S	pH independent
Types include anionic, non-ionic, and amphoteric			OM inactivation slow
Peroxides			
Peracetic acid and hydrogen peroxide	S	S	OM inactivated Corrosive

S: sensitive; M: moderately sensitive; OM: organic matter.

a grower should be familiar with the safety data (toxicity of the product, requirements for protective clothing, respirator filter specifications, the correct handling of the concentrate), potential environmental hazards, the most effective way to use the product for the purpose required, and to how to dispose of any excess. Storage can also be an important factor and some materials, for instance those containing chlorine, need to be stored at low temperatures and in the dark.

Choice of disinfectant

A number of factors must be taken into account when choosing the most appropriate disinfectant. Because of the different environments within a farm, no one disinfectant may be suitable for all uses. The requirements for a disinfectant used in foot-dips, and for dipping harvesting equipment, are not the same as those for washing down at the end of a crop. Whichever disinfectant is chosen, the instructions on its safe use should be read carefully before starting, and the appropriate protective clothing worn at all times.

Phenolic disinfectants are widely used for washing down growing rooms and compost preparation areas. They work well in the presence of compost and casing and remain effective in tanks for up to a week. But they are not suitable if run-off from the farm is likely to drain into a river or drainage channel. Very small quantities of phenolics in fresh water can have serious ecological effects, and water authorities are generally very anxious to prevent such pollution. If it is necessary to add a wetting agent to a phenolic disinfectant, it is important to choose an anionic product. An alternative in these situations is glutaraldehyde, particularly in QAC mixtures. These work well in the presence of compost and casing and are effective at high pH (alkaline).

For foot-dips in open situations, phenolics are widely used. Where the harvesters are likely to be in regular contact with the disinfectant, a QAC or an amphoteric biocide may be preferred. The latter are all environmentally friendly and can be used with safety within a cropping house.

Chlorine-containing disinfectants are often used to wash concrete surfaces. They appear to be able to soften the encrusted compost, enabling the power wash to achieve a clean finish.

Formalin (38-40% formaldehye), is effective in the presence of organic material and it is active in the vapour phase. Because of possible health risks its use is not permitted in some countries and local regulations must be consulted. Full protective clothing and a filtered aspirated helmet, must be worn. Treated areas must not be entered until the concentration of fumes has dropped to a safe level.

The uses of Formalin include casing treatment, fumigation of rooms and empty trays and, sometimes a spray onto the surface of paper covering a spawn-run. For casing treatment a 1% solution has been recommended using 25 litres/m^3. Even slight overdosing can be detrimental so this treatment should only be considered as a last resort. Treated casing must be stored in a protected area until the fumes have been eliminated.

Fumigation, often with a fogging machine (pressure jet), is effective for buildings but they must be reasonably air-tight in order to maintain a lethal concentration for about 1 hour. The treatment works best at 15°C or above. Free water in the treated area rapidly absorbs the fumes and reduces the concentration in the air. In the UK the maximum permissible rate of use is 4ml /m^3. This rate may vary from country to country and local regulations must always be consulted on the rate and conditions of use.

A 0.25% solution is sometimes used to spray the paper covering spawn-runs to prevent contamination of the compost from spores on the paper when the paper is removed.

Where a farm does not have steam available for cook-out and must rely on disinfection with chemicals at crop termination, the choice of a suitable disinfectant is particularly important. At crop termination, the crop should be sprayed with disinfectant, and if the spray triggers browning of any mushrooms remaining on the bed, the operator will be able to see the distribution of the spray. Areas unsprayed, and therefore not treated, are rapidly identified, and should be treated within a short time of the initial treatment. Both the phenolics and the glutaraldehyde mixtures are suitable for this purpose.

Cultural and environmental control

Pest and disease management programmes include both cultural and environmental aspects. Basic cultural factors such as the moisture content of the air or the compost, the management of watering, the temperatures of the compost or of the growing rooms, all play a very important part in the effectiveness of the overall pest and pathogen management strategy. The term environmental control is used in this context to encompass not only the environment in which the crop is grown, from spawning to harvesting, but also the environment of the compost during preparation.

Composting

The raw ingredients of compost may be contaminated with pests and pathogens, as well as antagonistic or competitive moulds. Some of these may be killed in phase I composting, but it is phase II (peak heating or pasteurizing) that should eliminate them. Very high levels of contamination in phase I can make eradication in phase II more difficult. Such levels can occur when soil or soil water drains onto phase I yards, or soil is brought in on straw. Soil contamination has been implicated in the appearance on farms of Pythium black compost and False truffle (**45**).

Inefficient preparation of phase II compost will have very serious consequences for successful cropping. Pasteurization at approximately 60°C for several hours is adequate to kill pests, pathogens, and most compost moulds but may not be reached throughout the compost.

Gaseous ammonia is a powerful disinfectant and is produced during composting. High levels of gaseous ammonia together with lethal temperatures in phase II are sufficient to eliminate pathogens, moulds, and pests. Generally, in order to get the best possible kill of unwanted organisms, a gaseous ammonia concentration in excess of 450 ppm measured three hours, into phase II, is very effective. Concentrations range from 200 to 1500 ppm in successful phase II treatments. However, lower levels, especially when they are as low as 200–250 ppm, have been associated with compost problems such as Trichoderma compost mould and False truffle.

Loss of temperature control during spawn-running is more likely when the process takes place in shelves, trays, blocks, or bags, than in bulk spawn-running tunnels (phase III). Compost temperature in excess of 32°C result in the death of mushroom mycelium. If temperatures reach 28–30°C they will favour *Trichoderma aggressivum*, some *Penicillium* species, *Pythium oligandron*, and *Diehliomyces microsporus*, all of which have optimum growth temperatures of 28°C.

45 A phase I area surrounded by soil banks which can lead to contamination of the compost with soil-borne pathogens. A serious Pythium black compost problem occurred on this farm.

Cropping

Water is perhaps the major environmental factor affecting disease development during cropping. The interacting effects of other environmental conditions such as air speed, temperature, carbon dioxide concentration, and relative humidity (RH), are difficult to separate.

Diseases which depend upon water or very high relative humidity for their development include the spotting diseases of the mushroom cap, particularly bacterial blotch and fungal diseases such as Trichoderma spot, Cobweb spot, Verticillium spot, and Aphanocladium. All of them can be controlled, or at least reduced, if the conditions in the crop are not suitable for their development. In this respect the management of the environment is vitally important, and for this, effective air conditioning equipment is essential. In order to achieve satisfactory quality and yield, a balance must be achieved between sufficient water for cropping, and enough evaporation to prevent persistent moisture that favours the germination of spores or the growth of bacteria. Loss of water from the crop surface is an essential, not only for good cropping but also for disease control. Evaporation from mushrooms takes place at about the same rate as it does from an open water surface. Drying of mushrooms is achieved on most farms by the manipulation of air temperature, relative humidity (or on some farms the absolute humidity), and the speed of the air flow over the bed surface.

Relative humidity is the extent (%) of air saturation at a given temperature, whereas absolute humidity is the total weight of water that can be held in the air at a given temperature (g/kg). Absolute humidity control can only be used successfully if the temperature is very precisely controllable.

The relationship of relative and absolute humidities is illustrated with the following example. At 18°C and at a RH of 80% the air holds 10.4 g of water per kg, whereas at 17°C and at 80% RH the same amount of air holds 9.8 g of water. Thus at the lower temperature the air can take up less water before it becomes saturated. According to Dutch sources, the water content of the air should not be above 11 g per kg of air at temperatures of 17–19°C, with a deficit of at least 2 g/kg (1 kg of air takes up 0.8 m^3 at standard temperature and pressure). This means, at 18°C with a deficit of 2.0 g/kg, the RH is 85%, whereas at 17°C the RH is 81%. If the relative humidity is set at 85%, the capacity of the air to take up water at 17°C is less than at 18°C and may not be adequate. Conversely, if the evaporating power of the air is too high, the mushrooms become scaly.

Research work has shown that at cropping temperatures, 85% RH and an air speed of less than 2 cm/s favours blotch development, and a speed in excess of 5 cm/s favours scaling. At 90% RH the figures are 3 and 8 cm/s respectively. Therefore, air speeds between these extremes at normal cropping temperatures should give no scaling or blotch (**46**). Farm surveys indicate that an evaporation rate of 0.8 mm of water per day, as measured by a Piché evaporimeter (**47**), is satisfactory. The Piché evaporimeter measures evaporation over a period of time, but not at any one moment and is therefore limited in its use. Most growers use set points for relative humidity and air temperature, relying upon their observations to avoid scaling and blotch.

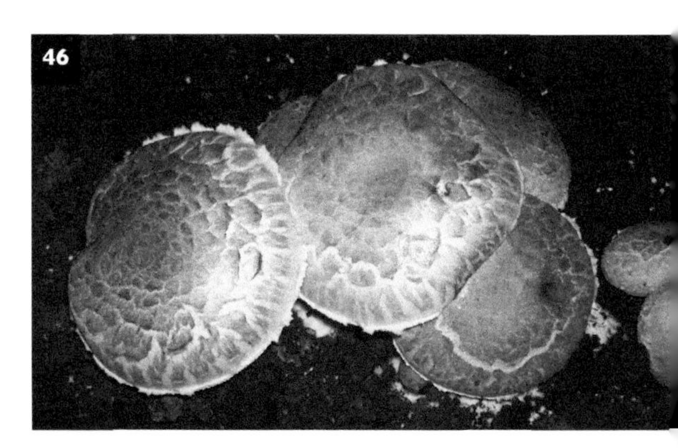

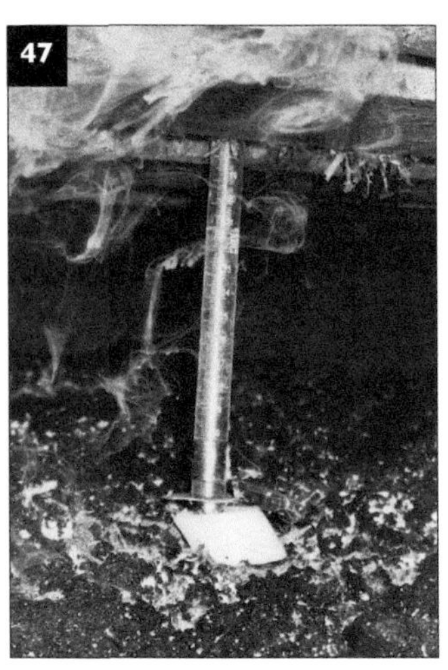

46 Feathering or scaling of a mushroom as a result of excessive air movement. Brown mushrooms are particularly sensitive.

47 A Piché evaporimeter with cold smoke demonstrating lack of air movement over the bed surface.

Action points during compost production and cropping

Phase I compost

- Bulk organic components must be checked for freedom from soil.
- Straw and other bulk materials must not be stored on a soil surface.
- Straw or other bulky materials with plant roots attached should be avoided.
- Water from surrounding land must not drain onto the phase I area.
- The phase I area must have a well maintained concrete surface so that it can be effectively power washed.
- Drain water that is reused should be checked for high concentrations of nutrients and toxic chemicals.
- Materials stored for use in phase I should be in such a position that fragments from them will not contaminate phase II or growing room areas.

Phase II compost

- Fresh air used to ventilate tunnels should have absolute filtration.
- Filters and ducts must be checked regularly for efficacy and leaks respectively.
- Bulk phase II compost must be very evenly loaded into tunnels.
- The plenum of tunnels should have organic debris removed regularly.
- Tunnels, including nets, should be cooked-out regularly.
- Shelves and trays used for the preparation of phase II *in situ* must be effectively cleaned between crops.
- The accuracy of the temperature recording equipment used during phase II must be checked regularly.
- The concentration of ammonia 3 hours into 'kill' should be at least 450 ppm.
- Contamination of compost by dust and debris at cool-down must be avoided.
- Phase II compost must be removed from the tunnel without the risk of contamination with dust or debris.

Phase III compost

- Compost must be filled at one end of the tunnel and removed from the other.
- Compost must be filled as evenly as possible.
- Absolute filtration must be used for the air that controls the compost temperature.
- The oxygen content of the compost atmosphere must not drop to below 16%.
- Equipment used to fill phase III should be dedicated to this process.
- Phase II spawning areas should be slightly positively pressurized.
- Workers should have dedicated overalls and boots which are only worn in the phase III area.
- Protective clothing should be cleaned daily.
- Boots should be disinfected every time the phase III area is entered.
- Nets should be steralized regularly.
- All equipment used to move spawned phase II into the tunnels must be disinfected before use and power washed afterwards.
- Similar procedures apply to equipment used to remove phase III compost from tunnels.

- Equipment used for blocking or bagging phase III should be regularly cleaned and disinfected.
- All transport equipment entering the loading area should pass through a disinfectant trough which disinfects the wheels.
- Tools and implements should be dedicated to phase III use only.
- Instruments recording the compost environment must be checked regularly and recalibrated if necessary.

Spawning and spawn-running for trays, shelves, bags, and blocks

- Air used in temperature control must be absolute-filtered.
- Spawn-running rooms must be cleaned, preferably by cook-out, between crops.
- Floors must be disinfected after cook-out and before spawn-running.
- Doors of spawn-runs must be kept closed at all times.
- Compost temperatures must be checked regularly.
- Spawn-running compost must be covered with paper.

Casing and casing run

- Casing ingredients must be stored in a clean area and kept free from contamination by air-borne spores and dust.
- Dust and debris must not contaminate the casing when it is being applied.
- Casing inoculum must be carefully and cleanly mixed into casing before application.
- Mushroom mycelium in compost, used as casing inoculum (cacing), must be carefully chosen from healthy spawn-run compost.
- Ruffling equipment used on casing must be disinfected before use and preferably at regular intervals during its use.
- Instruments recording the environment must be checked regularly.
- Excess casing should not be power washed from the floor of cropping rooms unless the floors are free from cracks.

Cropping

- Pickers must use clean overalls every day.
- Knives and other equipment used for harvesting must be regularly disinfected during the working day.
- Disinfectant troughs or pads must be placed in the doorways of cropping houses.
- Doors of cropping houses must be opened as little as possible.
- Disease removal teams must have full authority to do the job.
- Disease removal teams must have adequate light to allow thorough bed inspection.
- Pest and disease levels must be recorded.
- Disease removal teams must be able to recognize crop abnormalities and know how to deal with them.
- Pesticide use must be decided after taking into account current pest and disease levels.
- Crops should be inspected for disease before watering.
- Crops should be inspected for disease before harvesting.
- Generally the youngest crops should be inspected first and the oldest last, but this order can be changed according to the disease status of the crops.
- Pickers must never handle disease.
- Debris must be removed from the bed surface and the house after harvesting.
- Power washing the floor of a cropping house must be done with care to avoid splash onto the beds.
- Temperature and humidity readings should be checked regularly and instruments recalibrated if needed.

Genetic resistance

Although there is now known to be considerable variation in the genetic resistance of various species of *Agaricus* to the pests and pathogens of the mushroom crop, it has not so far been possible to use this variation. There is very little genetic difference between any of the commonly grown strains of *A. bisporus*, and they are all more or less equally susceptible to the known pests and pathogens of the crop. Some small variations have been reported, for instance brown strains appear to be less severely affected by Trichoderma compost mould than do white strains. When virus disease was first described, it was found that a break crop of *A. bitorquis* was useful as it escaped infection, possibly because it did not anastomose with *A. bisporus*, but it was difficult to grow and also to market. Also some of the rough white strains of *A. bisporus* (so-called virus breakers) appeared to be less susceptible, and were used to give a break from the smooth white types. Examination of collections of wild strains of *A. bisporus* shows variation in susceptibility to Verticillium, and to Bacterial blotch. Within the next decade, it is to be hoped that genetic manipulation may result in resistance from wild species being incorporated into the cultivated mushroom. At present this appears to be some way off and, even if successful, such genetically manipulated strains may not be acceptable to the consumer.

Organic production

It is possible to produce mushroom crops without the use of pesticides. Organic producers in many countries do this, although they represent only a small proportion of the total market. It is not part of the scope of this book to describe the details of organic production, but a key element of successful cropping is the control of pests and diseases. Some assistance is available for pest control in the form of commercial preparations of nematodes and bacteria. These are particularly effective for the control of sciarids. A fungicide, Trilogy, derived from the neem tree (*Azadirachta indica*) has recently been recommended in the USA for the control of Cobweb and Dry bubble diseases. Products from this tree are well known for their insecticidal and properties (Amazin, Azatin) and may have relevance in organic production. No such products are available for the control of fungal pathogens. In the absence of disinfectants and pesticides, the organic grower must rely almost entirely on good hygiene. Crop termination and tray cleaning with heat, are an essential part of the process. Power washing with water, if done very thoroughly, can take the place of disinfection. By growing only two flushes, pest and disease levels are generally contained, and epidemic development is prevented. Genetic resistance would be a very valuable additional tool for the organic producer, but pest- and disease-resistant strains are not yet available.

CHAPTER 4

Fungal Diseases

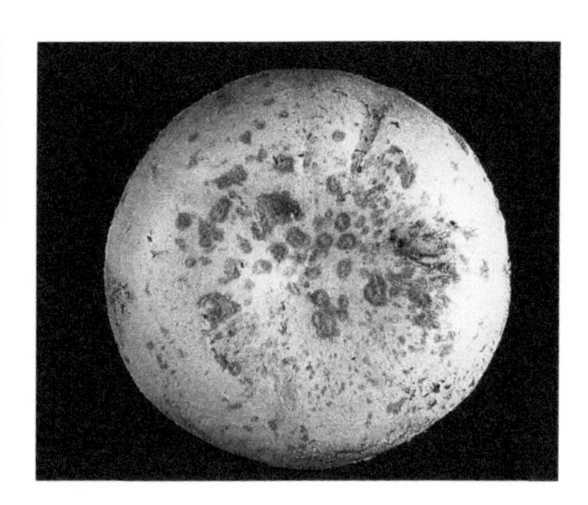

Introduction

Fungi are the most important group of mushroom pathogens. Fungal diseases can be found to some extent at any time in most countries but their quantity and severity vary. For instance, in the UK *Mycogone perniciosa* was the most common fungal pathogen in the 1950–60 period, followed by *Verticillium fungicola* var. *fungicola* from the late 1960s to 1980. In the mid 1980s to the early 1990s, *Trichoderma aggressivum* caused serious crop losses, and in the mid 1990s *Cladobotryum dendroides*, a pathogen previously rarely seen causing damage, was frequently epidemic. More recently, Verticillium has again become predominant, although all the others occur from time to time. In the USA, an epidemic of Trichoderma compost mould followed the outbreaks in Europe, and more recently Cobweb disease has caused serious problems on some farms. Similar patterns have been seen in Australia. With the international exchange of materials and greater uniformity of methods, it is possible that the occurrence of mushroom fungal pathogens may become more uniform internationally. There are indications of this already happening with the widespread use of black peat and the occurrence of Cobweb disease in a number of countries.

Most fungal pathogens can be effectively controlled by careful farm management, and in particular, by attention to hygiene (*see* Chapter 3). Effective control requires a good knowledge of their biology and should never rely too heavily on the use of fungicides. Routine use may encourage the development of fungicide-resistant strains as well as contributing to environmental pollution, and being an unnecessary cost. With the decreasing numbers of fungicides available, their correct use is very important. Organic producers rely entirely on good hygiene and modifications in growing methods for disease control and demonstrate that it is possible to produce good quality mushrooms without the use of pesticides.

At present, the option of biological control or genetic resistance is not available for disease control.

Wet bubble or Mycogone

This common disease is caused by the fungus *Mycogone perniciosa*, although it generally does not cause major crop loss. It is most serious when it develops early in a crop and is ignored. A second species, *Mycogone rosea*, has also been reported but is very uncommon. The horse mushroom (*Agaricus arvensis*) is commonly affected by *M. perniciosa* in the field (48, 49), although isolates from this source have been shown to be far less aggressive on *Agaricus bisporus*.

48 *Agaricus arvensis*, the horse mushroom, naturally infected with *Mycogone perniciosa* and showing the characteristic distortion symptom.

49 A sector of the gills of *Agaricus arvensis*, the field horse mushroom, affected by *Mycogone perniciosa*.

Symptoms

The most characteristic symptom of Wet bubble is the development of distorted undifferentiated mushroom tissue, which is initially white and fluffy but becomes brown as it ages and decays (**50**, **51**). These are the 'bubbles' which give the disease its common name. Early scientific descriptions called them sclerodermoid masses. They are often very large and can be 10 cm or more across. They result from infection at a very early stage in the development of the mushroom, which prevents differentiation into the normal tissues of the stalk, cap and gills. Growth is in an uncontrolled fashion, and a distorted mass of mushroom tissue results. Sometimes there is partial differentiation and partly formed caps may show wart-like protuberances on their surfaces (*see* **62**). From the time of infection to the appearance of the distortion symptom may be 10–14 days. Small amber to dark brown drops of liquid are sometimes present on the surface of the undifferentiated tissue, especially in conditions of very high humidity. When cut across, the affected tissue is generally dark in colour and wet in appearance. It is the wet decay as well as the shape of affected mushroom tissue that gives the disease its Wet bubble name. In drier conditions the distorted masses remain dry, and may then be confused with similar symptoms caused by *Verticillium fungicola* Dry bubble disease, although the distorted masses are generally larger.

In addition to distortion, *M. perniciosa* may produce small fluffy white patches of mycelium on the surface of casing, following the infection of a developing mushroom below the casing surface (**52**). This white mycelial growth is usually very distinct from mushroom mycelium and turns brown as it ages. With one isolate from China, surface mycelium was pale yellow-brown in colour (**53**).

When differentiated mushrooms are attacked, the stalk is colonized, with a brown streak reaching the cap, and a white sector or pie-slice shaped portion of colonized gills. The gills are stunted as well as being covered by the characteristic white mycelium of the pathogen (**54**). Sometimes only the base of the stalk may be affected. It has an internal brown discolouration, and eventually white fluffy mycelial growth of the pathogen develops. Affected stem bases, if left in the bed after cutting the mushrooms, can be a significant source of inoculum.

Cap spotting caused by this pathogen has not been found in crops, but has been induced in experiments when mushrooms were inoculated with isolates collected from naturally infected wild *Agaricus arvensis*, and to a lesser extent with an isolate from *A. bisporus* originating from China. The spots were dark brown in colour and well defined (**55**). A diffuse metallic brown staining of the caps has also been induced in experiments (**56**). Although *A. arvensis* and *A. bitorquis* are grown commercially, there have been no reports of their being affected by Wet bubble.

50 A mass of undifferentiated tissue of *Agaricus bisporus* resulting from infection by *Mycogone perniciosa*. The amber drops of liquid give this disease its common name of wet bubble.

51 A large mass of undifferentiated tissue of *Agaricus bisporus* affected by *Mycogone perniciosa*. Note the size of the diseased tissue in comparison with the surrounding healthy mushrooms.

52 White fluffy patches of mycelium of *Mycogone perniciosa* which occur on the surface of the casing when a developing mushroom, under the casing, is infected.

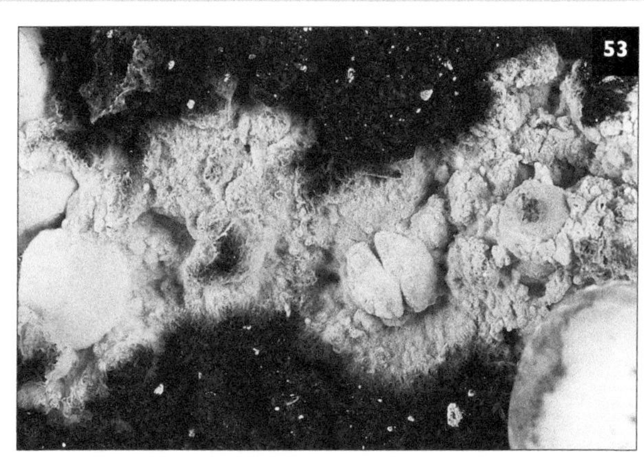

53 Yellow-brown mycelium of *Mycogone perniciosa* on the casing surface. This isolate of the pathogen originated in China.

54 A large portion of the gill tissue of *Agaricus bisporus* colonized by *Mycogone perniciosa.* This symptom occurs when the mushroom is infected well after differentiation of the stalk and cap.

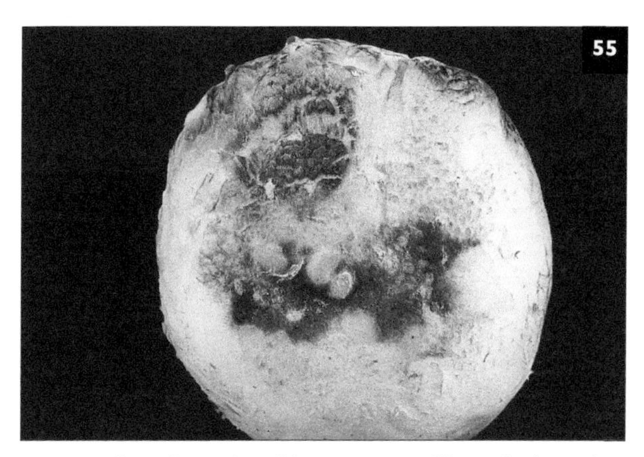

55 Spotting of *Agaricus bisporus* caused by an isolate of *Mycogone perniciosa* from *Agaricus arvensis*.

56 Diffuse brown staining on the cap of *Agaricus bisporus* induced by *Mycogone perniciosa* in experiments. This symptom has not been recognized in commercial crops.

Most of the symptoms shown by Wet bubble are also shown by Dry bubble. But some of the symptoms are more common with one disease than with the other; *Table 4* summarizes the differences.

Mycogone rosea does not distort *A. bisporus*; spots occur on caps, and white mycelial sectors develop on gills (57, 58). The cap spots are circular and brown in colour and are often surrounded by a pale yellow halo. These symptoms have not been reported from commercial crops. In general this fungus appears to be less virulent than *M. perniciosa*.

The pathogen and disease development

The pathogen produces two spore forms, one a single-celled thin-walled and relatively short-lived conidium, and the other a two-celled aleuriospore (terminal chlamydospore) with a very thick-walled terminal cell. It is believed that aleuriospores can survive for very long periods, and there is circumstantial evidence suggesting that they remain viable for at least 3 years in relatively dry organic material (stored casing). The conidia are light and may be air-borne, although there is very little

TABLE 4 A comparison of the symptoms of Wet (*Mycogone perniciosa*) and Dry (*Verticillium fungicola*) bubble

Symptom	Frequency of occurrence	
	Wet bubble	Dry bubble
Large bubbles	****	*
Small bubbles	**	****
Amber droplets	***	*
Wart-like growths	*	***
Casing surface mycelium	**	–
Cap spots	*	***
Gill symptoms	**	–
Stalk symptoms	*	**

Key: **** Almost always present; *** Very common; ** Sometimes present; * Occurs very infrequently; – Not seen.

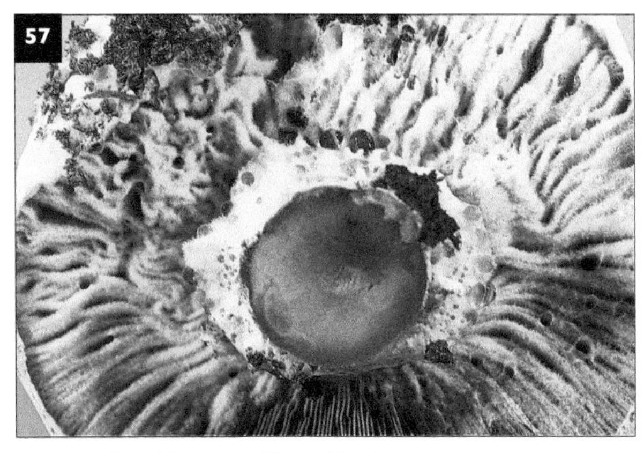

57 *Agaricus bisporus* affected by *Mycogone rosea*. Note that the gills are entirely colonized although affected sectors are also common. Amber droplets of liquid can be found on the affected tissue.

58 Brown spots on *Agaricus bisporus* caused by *Mycogone rosea*. Notice the yellow halo around the spots.

evidence that the pathogen spreads in this way. The aleuriospores and conidia are commonly spread by water splash. The fungus is found in soils. Outbreaks of the disease are said to follow soil movement on or near to the farm, or the use of soil-contaminated casing ingredients. The soil-borne fungus *Rhopalomyces elegans* has been reported to be a host of *M. perniciosa*.

The initial source of the pathogen on most farms is often traced to contaminated casing. Generally, symptoms in the first flush indicate contamination, either of the casing ingredients before they reach the farm, or on the farm during storage or mixing; compost is not a source. Spores may survive on surfaces or may be carried on crop debris, and in this way can contaminate crops. There is no evidence that *Mycogone* spores are spread by flies or mites or are air-borne.

When the pathogen is established in the crop, the main means of spread is by water splash and by excess water running off the beds. Spores on the floor can be readily redistributed back onto the bed surface by splash or in small droplets of water, especially if the floor is power washed. Pickers may also spread the pathogen on their hands, on tools, trays, and clothing, although, because the spores are not sticky, this is not an important means in contrast to the situation with *Verticillium*. There is little evidence that this pathogen will grow in the casing or compost although there are reports of growth alongside mushroom mycelium. The exact significance of this has still to be determined. In practice, it appears that diseased mushrooms result when mushroom initials develop very close to *Mycogone* spores. Exudates from developing mushrooms probably stimulate the germination of aleuriospores. Spores of the pathogen in the casing below the zone of mushroom initiation have been shown to have no effect and do not produce disease.

There is circumstantial evidence that yield loss occurs when this disease is severe, and the reduction is not solely the result of mushroom distortion. It seems likely that some mushroom primordia are destroyed well before they develop into recognizable mushrooms.

Control

One of the most important means of control is the elimination of the sources of the pathogen. Casing can be a source and it is particularly important to ensure that casing materials are clean and stored in an area where they will not become contaminated by debris or dust from cropping houses or drainage water or soil from nearby areas. Casing known or suspected to be contaminated can be heat treated (*see* p. 14).

Once the pathogen has become established in a crop, spread must be minimized. All affected mushrooms should be covered in salt. Very large bubbles can be removed using a polythene bag, paper towel, or plastic glove on the hand. Bags and gloves can then be inverted and tied for safe disposal. All evident disease must be removed before the crop is picked (*see* p. 41).

As watering is one of the most important means of spread of the pathogen, the crop should be watered only after all the diseased mushrooms have been removed.

Prochloraz manganese (Sporgon/Octave) gives a good control of Wet bubble, as does carbendazim (Bavistin) and the other benzimidazole fungicides (Benlate and Tecto). There are no records of *Mycogone* developing resistance to these fungicides except for one report of carbendazim resistance in Korea.

Disease control with carbendazim (Bavistin, Benlate) may fail because of the rapid breakdown of the fungicide in the casing (*see* Chapter 3, p. 52). In such cases it is worth changing to prochloraz manganese although in some circumstances this chemical may also degrade (*see* p. 52).

If the casing is contaminated, it can be treated at mixing and before use with 1% Formalin using 25 litres/m^3 (*see* p. 51). This treatment can be risky, because of the variable water content of different casing mixes. Overdosing can seriously affect cropping, and also appears to encourage *Trichoderma* spp. For these reasons, it is often considered as a last resort. A safer treatment, but one which is usually more difficult to apply, is the use of heat. The temperature of the casing is raised to 50°C, and maintained at this temperature for 30 minutes. Steam–air mixtures at 80°C have also been used.

Wet bubble or Mycogone action points

- Make a positive identification in order to apply the correct treatments.
- Observe strict hygiene at all times, paying particular attention to disease removal (*see* p. 41), picking sequence (*see* p. 43), and watering.
- Effective crop termination is essential (*see* p. 46).
- Avoid power washing the floors of cropping houses, especially if the floors are cracked.
- Make sure the casing is mixed and stored in a clean area which is, as far as possible, free from the dust and soil.
- When the risk of Mycogone is high, use carbendazim or prochloraz.
- Where control of Verticillium is also a consideration, prochloraz manganese should be used.

Dry bubble or Verticillium

This is a serious and common disease of mushrooms wherever they are grown. If left uncontrolled, Dry bubble can reduce farm income to the point where it is not possible to produce mushrooms economically. The disease is caused by the fungus *Verticillium fungicola* (syn. *V. malthousei)*; different varieties of this pathogen occur. In Europe the variety *fungicola* predominates. A second variety, *aleophilum*, appears to be more common in warmer countries and in the USA. *V. fungicola* var. *aleophilum* is also associated with crops of *A. bitorquis*. A closely related fungus, *V. psalliotae*, occurs infrequently and causes symptoms similar to those caused by the variety *aleophilum*.

Symptoms

The symptoms caused by *V. fungicola* var. *fungicola* are varied and depend upon the developmental stage of the mushroom at the time of infection. The distortion symptoms of both Wet and Dry bubble diseases are very similar (*see Table* 4).

At an early stage in mushroom development, small undifferentiated masses of tissue, generally up to 2 cm in diameter, occur (**59**). These can sometimes be larger, and may then be confused with those of Wet bubble (**60**). They may even ooze droplets of amber-coloured liquid. When affected at a later stage in their development, mushrooms are often

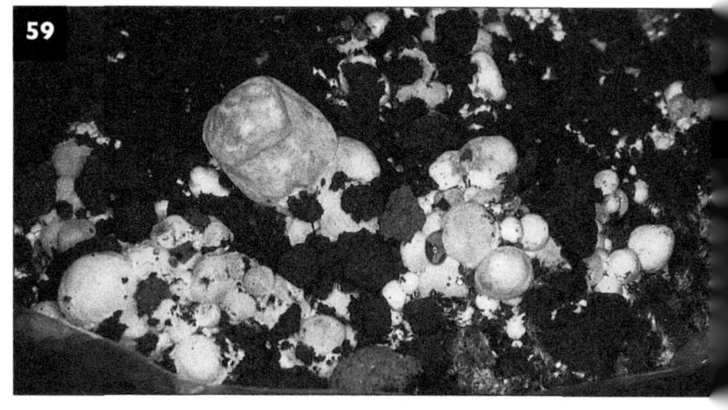

59 A cluster of small grey-brown undifferentiated masses of mushroom tissue (*Agaricus bisporus*) resulting from infection by *Verticillium fungicola* var. *fungicola*. Infection after differentiation has resulted in the partially formed mushrooms.

imperfectly formed with partially differentiated caps, or with distorted stalks and malformed caps (**61**). Such affected mushrooms are covered in a fine grey-white mycelial growth and, although discoloured, are generally dry in appearance. As they age, the distorted masses turn light brown in colour. Occasionally, almost fully differentiated mushrooms are affected, and these show small pimple or wart-like outgrowths on the top of the cap (**62**).

60 (**a**) Larger undifferentiated masses of mushroom tissue (*Agaricus bisporus*) following infection by *Verticillium fungicola* var. *fungicola.* Notice the droplets of liquid similar to those generally associated with Wet bubble. (**b**) Within a flush of mushrooms such relatively large distorted mushrooms are easy to see.

61 (**a**) Infection early in differentiation of the mushroom (*Agaricus bisporus*) results in a more mushroom-like mass and (**b**) distortion of a fully developed mushroom when infection occurs after the end of differentiation.

62 Small wart-like growths on the cap of affected mushrooms. Both *Verticillium* and *Mycogone* can cause this symptom on *Agaricus bisporus*, but it is far more commonly caused by *Verticillium*.

One of the commonest symptoms in an epidemic of the disease is cap spotting. Generally the spots are blue-grey (1–2 cm in diameter), turning brown as they age (**63**). Sometimes the spots are accompanied by a yellow or bluish-grey halo. A less common form of spotting is sometimes seen. The spots are darker brown with well defined edges and the centre of the spot is slightly sunken to give a shallow crater (**64**).

V. fungicola var. *aleophilum* does not always cause distortion, but does affect the cap, on which it produces a dark brown spot (**65**). These spots are very similar to those of Trichoderma cap spot, Aphanocladium spot, and Cobweb spot, and may only be distinguished by laboratory examination. The Verticillium dark brown spots may eventually become covered by a blue-grey 'bloom' as the pathogen produces spores.

V. psalliotae does not cause distorted mushrooms, but cap-spotting is common, and is indistinguishable from that caused by *V. fungicola* var. *aleophilum*.

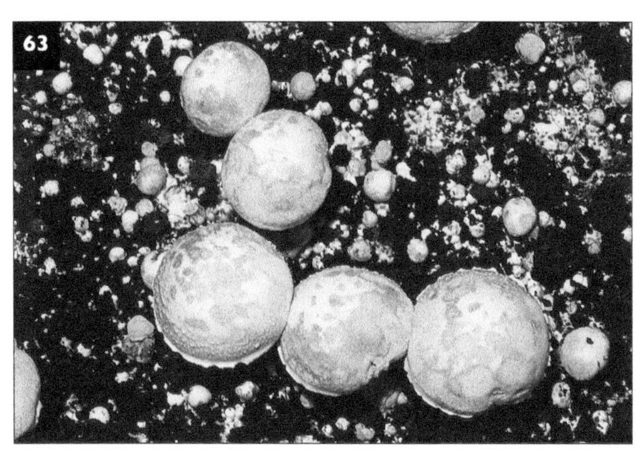

63 Cap spotting of *Agaricus bisporus* caused by *Verticillium fungicola* var. *fungicola*. The spots are characteristically light brown, sometimes with a blue-grey centre where the pathogen is producing spores.

The pathogen and disease development

Verticillium species produce thin-walled sticky conidiospores. There are no other known spore forms, but it is possible that thick-walled resting mycelium in a dormant state could enable the pathogen to survive for long periods. The most important initial source of this pathogen is not known. Although soil is often implicated as a source, the pathogen has only once been reported in soil. It is known to be an inhabitant of peat. The disease has not been recorded on field mushrooms. Some recent evidence indicates that it can grow on wood in moist conditions, and may persist from crop to crop or in the soil on wood. *Rhopalomyces elegans*, a soil-borne fungus, has been reported to be a host of *Verticillium psalliotae*.

Circumstantially, there is evidence that the disease appears following earth movements or building operations on or near to a farm.

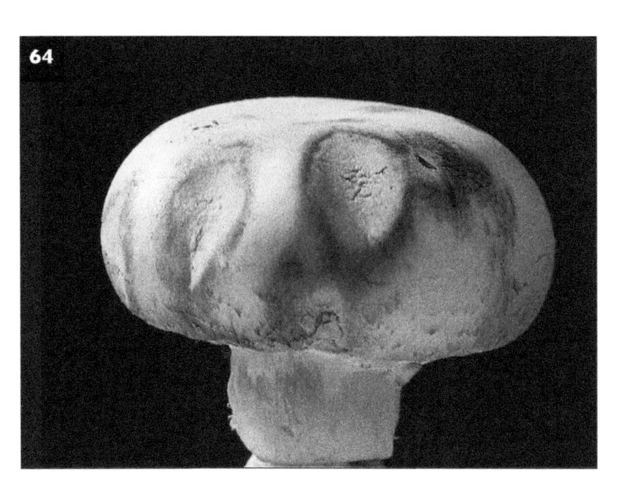

64 Sunken lesions with a dark brown margin on *Agaricus bisporus* resulting from infection by *Verticillium fungicola* var. *fungicola*. The conditions which result in this symptom are not known.

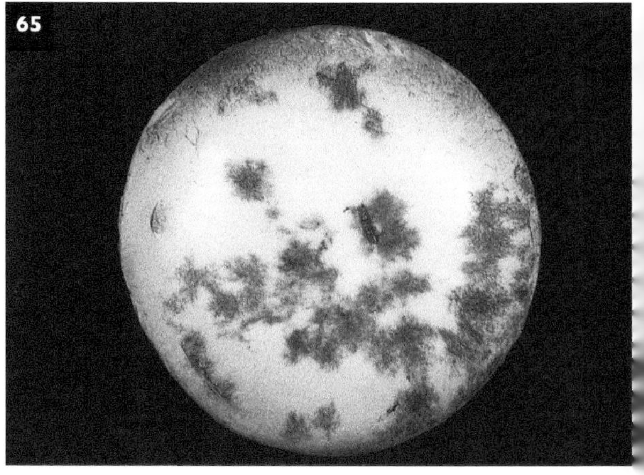

65 Very dark cap spotting of *Agaricus bisporus* caused by *Verticillium fungicola* var. *aleophilum*.

Returnable plastic containers cannot be ruled out as an initial source.

Unlike the spores of *M. perniciosa*, conidia of *Verticillium* are produced in clusters surrounded by sticky mucilage, and it is this mucilage which enables them to become attached to dust, flies, mites, debris, and pickers. Spores attached to dust particles can be a very important means of spread around a farm. It has recently been shown that dust on the floors of growing rooms may contains low concentrations of *Verticillium* spores, even though a good hygiene programme was in operation.

Spores on dust and debris from bed surfaces often accumulate in cracks and crevices on the floor of cropping houses. Power washing this from the cracks can often result in debris being splashed onto the casing surface. *Verticillium* spores can be transferred from the floor onto the beds in this way.

The distribution of spores on hands and clothes is also very important both within and between crops. Once hands are contaminated it is not easy to clean them, even using hot water and soap. Some mites (*Tyrophagus* spp.) are known to feed on the spores and mycelium of *V. fungicola* var. *fungicola*, and viable spores have been recovered from their faecal pellets. Such infested mites could be carried from house to house or even from farm to farm. Flies that walk on affected mushrooms are likely to pick-up spores, and in this respect both phorids and sciarid flies are a very important means of dissemination.

The pathogen can be introduced onto the farm on returnable containers, by personnel, or by vehicles.

Flies from an affected crop which are transported with harvested mushrooms from another farm can also be a serious source of the pathogen, particularly when containers are stored near to growing houses.

Although evidence to date suggests that spread by air-borne spores does not occur, in effect this happens when the spores adhere to small pieces of dust or debris.

Once the disease is established, there is no doubt that the farm is by far the most important local source of the pathogen. As with *Mycogone*, this pathogen does not grow through the casing or compost, although it is said to grow alongside mushroom mycelium. The exact significance, if any, of such growth, has still to be determined.

Disease is most severe when the inoculum level in surface layers of the casing (down to the zone of initiation) is high. As the pathogen does not grow through the casing, it is likely that disease develops only when mushrooms grow adjacent to spores. Inoculum in the lower layers of the casing or in the compost is not important. If inoculum concentrations are low, the chances of spores being near to developing mushrooms is reduced, and it may not be until the second or third flush that symptoms appear, even though the crop was contaminated before harvesting began. Therefore, when considering possible sources of inoculum, time of symptom expression and levels of disease in each flush must be taken into account.

Watering is an effective means of spore dispersal within the crop as excess water running off the casing carries spores to lower beds, or to the floor of the house. When the floor dries, air movement over its surface may cause debris carrying spores to become air-borne. Alternatively, they may be splashed onto the casing surface during watering.

V. fungicola var. *fungicola* grows fastest at 24°C. Although disease develops over a wide range of environmental conditions, the optimum temperature for development is 20–25°C. The time taken at 20°C from infection to the distortion symptom is about 10 days; for cap spotting it is 3–4 days. *V. fungicola* var. *aleophilum* and *V. psalliotae* grow fastest at 27°C. This may account for the infrequency of the latter two pathogens in crops of *A. bisporus* in northern Europe, and their greater frequency in warmer climates as well as in crops of *A. bitorquis*, which are grown at higher temperatures than *A. bisporus*.

If disease appears before harvesting the first flush, it is very important to consider the possible sources of the pathogen that have lead to such early contamination. *Verticillium* in the first flush often results in widespread disease by the third. Likely means of contamination include spread by flies or mites, by workers, contaminated casing, contaminated shelves/trays, and a contaminated growing room, in particular the dust and debris on the floor which may be water-splashed onto the crop. In shelf systems, contamination of the casing by air-borne dust at the time of application is always possible.

Control

A very careful and meticulous examination of all possible sources and means of spread of the pathogen is essential.

Returnable plastic containers are a possible initial source especially if they are taken directly into crops. It is important that they are thoroughly washed before reuse (*see* p. 44).

Casing ingredients can be a source as the pathogen has been found in peat. Contamination of casing can also occur after the ingredients have arrived on the farm. Disinfection of the storage area, and safeguards to prevent contamination from other parts of the farm, are essential. The process of casing is particularly vulnerable to air-borne inoculum. Every effort should be made to keep the casing area clean. A large foot-dip should be placed at the entrance to the casing area and contamination by drain water must be prevented. There should be designated equipment for handling and mixing casing, and this should not leave the casing area.

Casing can be heat-treated, but this is costly, and generally an unnecessary operation unless recycled spent compost is one of the casing ingredients (*see* p. 69).

Because of the easy distribution of spores on debris, by pickers, by mites, by flies, and probably by mice and rats, it is vitally important to salt all affected mushrooms at the earliest possible stage. Affected pinheads are easily missed, especially if lighting on the crop is inadequate. A disease removal team working systematically through the crops is essential (*see* p. 41). Every crop on a farm with Verticillium should have disease treated every day including weekends.

Spread can be prevented by the use of pots or similar physical barriers. It is important to push the pot well into the casing as far as the surface of the compost, in order to prevent the sideways drainage of water, which will distribute spores. A combination of pots and salt is more effective as it does not require the pots to be pushed into casing.

Best hygiene practice at harvesting is also vitally important. Pickers should only work in crops that have been recently checked and treated for disease. Also, watering must never be done before disease is treated (*see* p. 43). Watering and power washing floors of cropping houses can easily transfer *Verticillium* spores from the floor, where they may be present in cracks, onto the surface of the casing (*see* p. 54).

Flies must be controlled especially in the summer months, when populations are likely to be at their highest. Both sciarids and phorids may effectively spread *Verticillium* spores.

As levels of disease on the farm become increasingly severe, it is more and more difficult to achieve control by conventional means. In these circumstances, it may be necessary to terminate all crops after two flushes. If repeated for a complete cropping cycle, the inoculum level will be dramatically reduced.

The most important single process in the reduction of inoculum and the control of dry bubble is effective crop termination (*see* p. 46). Farms with persistent Dry bubble problems are often those that are unable to steam cook-out. Heating the spent crop to a temperature that kills the *Verticillium* on all surfaces, and also kills mushroom mycelium, guarantees that the growing room and the containers are free from the pathogen at the end of a crop. The thermal death-point for this pathogen and mushroom mycelium is quite low (between 45°C and 50°C). The temperature must be maintained for a long enough period to allow the heat to penetrate to all areas including into the wood used in tray and shelf construction. This is particularly important with *Verticillium* as it is able to grow on wood which may therefore be an important source.

If treatment cannot be done *in situ*, and the crop is transported from the growing room to a cook-out room, the chances of spread are increased, especially if live disease is present on the trays at the time of transportation (*see* p. 48).

After cook-out, empty trays must be stored in a place where they will not become re-contaminated. Efficacy of tray cleaning may also be improved by cooking-out empty trays (*see* p. 48), as well as by lining them with plastic before they are reused.

If thermal cook-out is not available, the crop must be sprayed-off. The empty house must be power washed, dried, and fumigated with formalin (4 ml/m^3) before refilling (*see* pp. 51–52). In this way the floor and structure are cleaned before the next crop arrives in the room.

Prochloraz manganese (Sporgon/Octave) is the only effective fungicide for the control of Dry bubble disease, although its use is not permitted in all countries. A two spray programme (where the label permits) is usually sufficient to give effective control. The fungicide can also be mixed in with the casing at the time of mixing the ingredients. This operation can be difficult with some casing mixes, particularly those containing a large proportion of fine-grade black peat. The method also has the disadvantage that half of the fungicide is not utilized, as it is below the zone of mushroom initiation.

During recent years, isolates of the pathogen have been found that are less sensitive to prochloraz manganese but are still controlled, although the length of time the fungicide protects the crop may be reduced. This can result in the appearance of disease early in the second flush. In an attempt to improve control, the fungicide has been applied with lower volumes of water which effectively increase the concentration of the active ingredient in the top layer of the casing. Although this may have some slight advantage in situations where the level of control is below normal, it may delay the crop slightly or reduce yields. There are no records of total failure of control because of fungicide resistance associated with prochloraz manganese, or of the rapid and total breakdown of the fungicide in the casing, although degradation has been found to occur in the spray tank (*see* p. 52).

Almost all populations of *V. fungicola* var. *fungicola* are resistant to the benzimidazole fungicides (Benlate, Bavistin, Tecto, Topsin M); strains resistant to one of the benzimidazole fungicides are also resistant to the others. For this reason benomyl, carbendazim, and thiabendazole are not worth considering. This may not be the case with all the varieties of the pathogen, as there is circumstantial evidence that strains of the variety *aleophilum* may be controlled, at least to some extent, by thiabendazole. In this case accurate identification of the pathogen is essential. Chlorothalonil (Daconil) is an alternative fungicide, but is only moderately effective and then only when disease levels are low. Recently a natural product, Trilogy, has been recommended in the USA (see p. 61).

Evenness of application of fungicides is very important, particularly if the pathogen population is less sensitive to prochloraz manganese. The distribution of the spray should be checked at intervals (*see* p. 51). With tray-growing, a spray boom on the casing line gives the most even application including around the tray legs which can be difficult to spray when the trays are in position.

Prochloraz manganese should not be used as a routine treatment, and should not be used when Verticillium levels are extremely low or the disease is absent. Low risk of Verticillium is often associated with small fly populations.

Although none of the commercially available spawn is resistant to Dry bubble, some differences in susceptibility have been found in Asian strains as well as in some strains in wild collections.

Action points

- Obtain an accurate identification to institute the most effective control.
- Observe strict hygiene throughout the farm (Chapter 3).
- Pay particular attention to crop termination and harvesting hygiene (*see* pp. 46 and 43).
- Make sure casing ingredients are stored and mixed in a clean area.
- Pay strict attention at casing to possible contamination by dust, especially in shelf systems.
- Do not power wash debris out of cracks and crevices in the floor of cropping houses after casing.
- Control flies and mites.
- Remove or treat all affected mushrooms before picking and watering (*see* p. 41).
- When disease risk is high, apply prochloraz manganese.
- Check for fungicide insensitivity and/or fungicide distribution if disease control is not as good as expected.
- Check for fungicide degradation (see p. 52) and thoroughly clean the spray tank if this is shown to occur.
- Terminate severely affected crops early by either cooking-out or the use of effective disinfectants.
- Use a two-flush strategy when disease levels in the third flush are very high.
- Line trays with polythene.

Cobweb or Dactylium

Cobweb disease can be caused by a number of different but related fungi. *Cladobotryum dendroides* (syn. *Dactylium dendroides*) which is the conidial state of *Hypomyces rosellus* has historically been considered to be the commonest cause. *Cladobotryum mycophilum*, the conidial state of *Hypomyces odoratus,* is also commonly found in Europe, North America and South Africa. *C. mycophilum* produces a characteristic odour on agar, which has been described as being similar to that of turpentine. A form of the pathogen showing some characteristics of *C. dendroides,* but genetically closer to *C. mycophilum*, has been found in the UK and has been referred to as *C. mycophilum* type 2. So far, all isolates of this form are resistant to thiabendazole fungicides. *Hypomyces aurantius* (stat. conid. *Cladobotryum varium* syn. *C. variospermum*), *Cladobotryum multiseptatum*, and *Cladobotryum verticillatum*, have all been recorded as causes of Cobweb disease, but there is no information on their incidence or importance.

In recent years, Cobweb disease has become common and a serious cause of crop loss not only in Europe but also in the USA and Australia. Its increase has coincided with changes in cultural techniques, in particular casing type and watering. Control with fungicides has become increasingly more difficult because of the development of resistance to the benzimidazole fungicides, and a decreased sensitivity of some forms to prochloraz manganese.

Symptoms

None of the various symptoms are associated with any one particular species of *Cladobotryum*. It is possible that all of the species involved do not produce all of the symptoms, which could account for some of the variation in appearance with locality.

One of the main symptoms of Cobweb disease is the cobweb-like growth of mycelium over the surface of mushrooms, hence the common name. The colonized surface turns pale brown (**66**). This discolouration, together with the presence of the off-white mycelium of the pathogen, is very diagnostic (**67**). Cobweb mycelium is able to grow over the surface of the casing in the absence of mushroom fruiting bodies, but not in the absence of mushroom mycelium. Patches of dense white powdery mycelium, sometimes circular in shape, may appear on the surface of the casing (**68**). Many spores of the pathogen are produced from this mycelium. Occasionally when left on the bed, diseased mushrooms turn a red or yellow colour (**69**). Affected mushrooms eventually turn brown or black and rot. Cap spotting also occurs and can cause large crop losses. There are two types of spot. The commonest has dark brown spots characterized by a poorly defined edge. They are very similar in appearance to those caused by *Trichoderma aggressivum* (**70**). The spots develop in 3–4 days from spore germination. Extensive spotting of a flush can occur when inoculum concentrations are high and conditions for spore germination favourable. Spots can also develop after harvest. In the second form of spotting, more or less circular spots, light brown in colour, develop. The

66 Mycelium of *Cladobotryum dendroides*, Cobweb, on the surface of the casing and attacking developing mushrooms (*Agaricus bisporus*).

67 Cobweb mycelium discolouring mushrooms (*Agaricus bisporus*) as it colonizes their surfaces. The mycelial strands hanging from the affected mushrooms give the disease its common name as they resemble cobwebs.

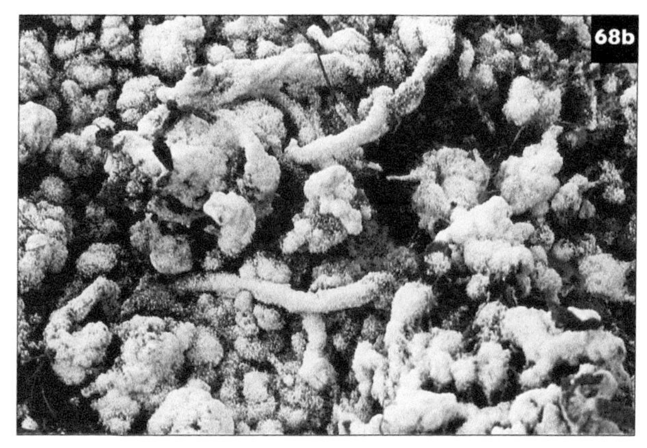

68 (a) Dense white circular patches of Cobweb mycelium sometimes develop on the casing surface. When seen close up, the mycelium can be seen to be powdery and is a major source of the spores of the pathogen (b).

69 Pink-coloured mushroom tissue in a Cobweb patch.

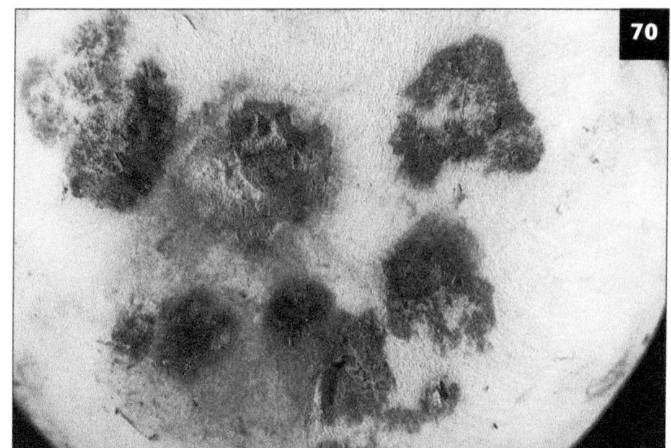

70 Spots on a mushroom from a crop (*Agaricus bisporus*) with a severe attack of Cobweb. These spots are very similar to those of both Aphanocladium and Trichoderma.

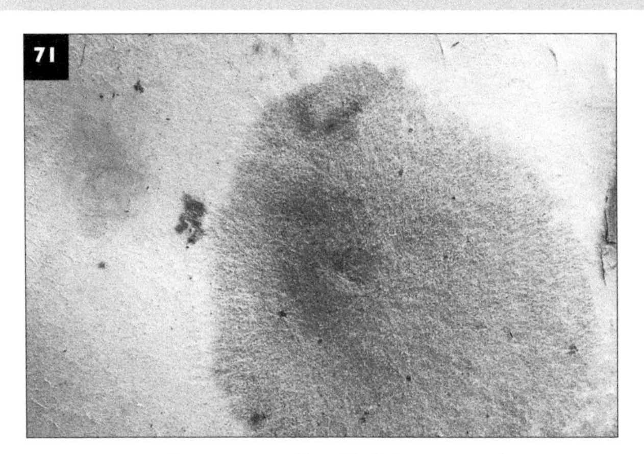

71 An unusual spot caused by *Cladobotryum*. These spots increase in size as the pathogen colonizes the surface of the mushroom (*Agaricus bisporus*).

72 A container of Cobweb-affected mushrooms which have been stored for a few days in a warm place. It only needs one or two affected mushrooms for the whole quantity to be colonized in favourable conditions.

spots appear to grow from a point source and radiate out until a large proportion of the cap is affected (71). Extensive colonization of harvested mushrooms may also occur after harvest when the crop is stored in damp conditions (72).

The pathogen and disease development

The pathogen *(C. dendroides)* grows best at 25°C, at 90% RH and pH 5–6, but sporulation is most abundant at 25°C and 95% RH. Conidia are the only commonly occurring spore form of this pathogen, although there are research records of the occurrence of microsclerotia in laboratory cultures. It is not known whether or not these are significant in practice. The spores are relatively large and multicellular, but are very readily dispersed in air. Any physical disturbance of the air near to a patch of Cobweb will result in spores becoming air-borne. This is the main way in which the pathogen is distributed on farms, but perhaps also between farms. Air-borne spores may contaminate returnable plastic containers taken into an affected crop and, in this way, the pathogen could be spread from farm to farm.

Even the operation of covering diseased mushrooms with salt or a pot can result in significant spore dispersal. The spores can also be spread by water splash (which also makes them air-borne), in drainage water, and by pickers. In recent epidemics flies, in particular sciarids, have been implicated in spore spread, but as the spores are not sticky, it seems unlikely that flies or even pickers are a significant means of dispersal. In addition to these mechanisms, the pathogen can be distributed as small fragments of mycelium present on the surface of mushrooms or casing. Small pieces may be carried in air currents or be accidentally spread during harvesting and be part of the dust.

The pathogen has been found in soil and is known to attack some species of wild mushrooms. Possible initial sources include soil, contaminated casing, air-borne spores and/or mycelial fragments, contaminated dust and debris, flies, and contaminated workers. Harvested mushrooms that are moved between farms for packing can also be a very important initial source, as well as the returnable plastic containers they are in.

Most commonly, the first symptoms of the disease are seen in the later flushes, but once established on the farm, they may appear at any stage. Experimentally, it has been shown that mycelial fragments applied to casing result in the rapid development of Cobweb symptoms, whereas even large spore loads may not immediately induce disease. Spores quickly cause cap spotting. Symptoms in the third flush may therefore have resulted from spore contamination of the casing either before or shortly after it was applied. Unlike *Verticillium* and *Mycogone*, this pathogen is capable of growth on or through casing. In this respect it is possible that contaminated compost can also be a source.

Disease development, whether it is the Cobweb

symptom or spotting, is considerably influenced by the presence of water or very high humidities. When water does not evaporate from the casing or the surface of mushrooms, Cobweb is likely to be severe. Sometimes the slope of a growing room is sufficient to result in humidity differences along its length, with Cobweb developing in the colder wetter parts.

Control

The use of wetter casing and the development of fungicide resistance have made this disease much more difficult to control. Surface wetness and a non-evaporating casing will encourage disease development. In order to help prevent epidemic development, the environment must be controlled to give continuous evaporation (*see* p. 58).

Effective steam cook-out is by far the best way to control Cobweb. Fresh spores are reported to be killed by a 30-minute treatment at 45°C, but are able to withstand much higher temperatures, even as high as 100°C when dry. Similarly fresh mycelium is killed at 40°C when this temperature is maintained for 15 minutes, but dry mycelium withstands 70°C for 15 minutes. On farms where it is not possible to achieve higher temperatures, cook-out at 50°C can eliminate many of the spores.

Spore dispersal resulting from watering-affected areas of crop, or careless application of salt to affected patches, is considered to be one of most important factors in recent outbreaks of the disease. Air-borne spores can contaminate any surface. For instance, returnable plastic containers used in an affected crop can easily become contaminated and must be cleaned before reuse (*see* p. 44). The same applies to any equipment or implements.

In order to minimize air-borne spores, affected mushrooms and areas of casing should be very carefully covered with a damp paper towel which is then covered with salt, starting at the outside of the patch and working inwards (*see* p. 42 and **34**).

It is important to examine the layout of the farm to check whether it is possible for spores to move from an affected crop to a healthy one. Some form of filtration on incoming and outgoing air minimizes spread. Absolute filtration should not be required for growing rooms, but filters that will remove most, if not all, of the air-borne spores, are needed. A 5-micron filter is usually adequate for this job. During picking of affected crops, the doors of houses should be kept closed as much as possible to prevent air moving from affected crops to nearby healthy ones. Positive air pressure used in cropping rooms can aid spore spread, especially at harvesting when the doors are open.

At crop termination, the spores may become air-borne and can then be transferred to a new crop (*see* crop termination, pp. 46–48). This is particularly the case during spray-off. If chemically terminated crops are taken through the farm it is vitally important to make certain that there are no untreated Cobweb patches on the cropping surface.

Before applying fungicides, it is important to know the sensitivity of the pathogen affecting the crop. Of the fungicides available, prochloraz manganese (Sporgon/Octave) and the benzimidazoles (Bavistin, Benlate, and Tecto) are effective, but not on all isolates. Some strains of the pathogen are moderately insensitive to prochloraz. Similarly, many strains are resistant to thiabendazole while being sensitive to benomyl or carbendazim. Some are resistant to all the benzimidazoles.

Chlorothalonil (Bravo) gives some control of all isolates but is not as effective as the other fungicides. Recently a natural product, Trilogy, has been recommended in the USA (see p. 61). Generally, the choice of fungicides is between prochloraz manganese and carbendazim.

Action points

- Cover affected mushrooms and areas of casing very carefully as soon as they appear, using damp paper towels and salt (*see* **34**).
- Never water or handle untreated areas of disease.
- Check harvesting procedures to minimize the chance of transfer of air-borne spore and mycelial fragments from one crop to another (*see* pp. 43–44).
- Do not harvest crops that have not been inspected and treated.

Contined overleaf

Cobweb or Dactylium action points (*continued*)

- Fit air filters to input and exhaust ducts in cropping rooms.
- Test for fungicide sensitivity.
- Apply the appropriate fungicide.
- It is important to prevent long periods of cap wetness and to avoid poor evaporation from the casing surface (*see* p. 58).
- Check all aspects of crop termination to eliminate inoculum at the end of each crop (*see* pp. 46–48).
- All returnable plastic containers must be cleaned before being taken into a crop.

False truffle

The False truffle fungus (*Diehliomyces microsporus* syn. *Pseudobalsamia microspora*) not only competes in the compost for food and space, but is believed to attack mushroom mycelium and cause mycelial death. There are very few recorded pathogens of mushroom mycelium and in this respect it is unusual. It is sporadic in occurrence and less common than it was, but when it occurs it can persist and be difficult to eradicate.

Symptoms, the pathogen, and disease development

D. microsporus is a common soil dweller and severe outbreaks on farms often follow operations which involve soil movement. When soil was used as casing, False truffle was a much commoner problem. Where *Diehliomyces* mycelium grows, the mushroom mycelium disappears, and the compost is black, often wet, and is said to have a characteristic chlorine-like smell. For severe crop loss to occur, False truffle must be present in the compost at or during spawn-running, and then yield reductions can be as high as 75%.

The mycelium of *D. microsporus* is initially white, becoming cream to pale pink in colour, and it often grows in dense cotton-wool like wefts (**73**). Initially it can be virtually impossible to distinguish from the mycelium of the mushroom (*see* Sectoring, p. 167), but becomes distinct with age as it changes in colour to red-brown (**74**). The False truffles (ascocarps) of the fungus form within the dense wefts of mycelium within 15–21 days, depending on the temperature, and are often the only indication of the presence of the mould, as they persist after the mycelium has disappeared.

False truffle is often first found when non-cropping areas of bed are carefully examined. Small creamy-brown coloured corrugated lumps of tissue (ascocarps) may be found in the compost of these areas. Ascocarps vary in length from 3 mm to 40 mm, and although initially light in colour, turn ochre yellow to red-brown as they age. The exterior corrugation on their surface gives them a characteristic appearance, often said to resemble a shelled walnut or the surface of a brain (hence the once common name 'calves' brains', **75**). The ascocarps most frequently form at the casing–compost interface (**76**), but also along the sides of trays or beds, or against the base of the bed, and

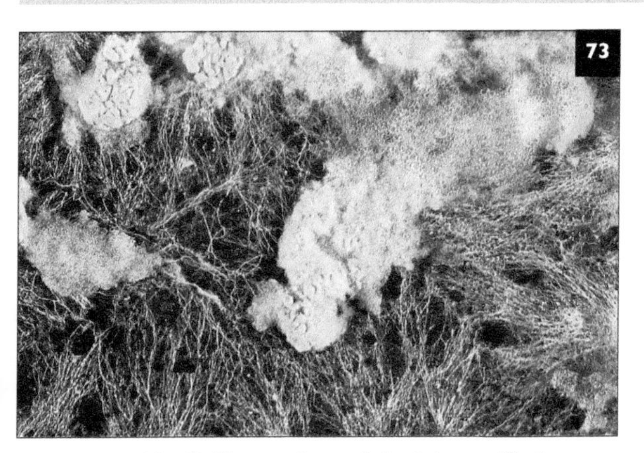

73 Dense white fluffy mycelium of the False truffle fungus in contrast to the whiter stringy mycelium of the mushroom.

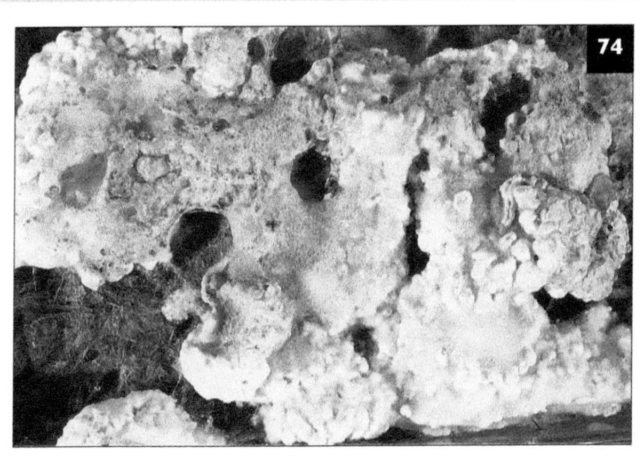

74 Dense growth of red-brown mycelium of the False truffle fungus. The colour intensity increases with age.

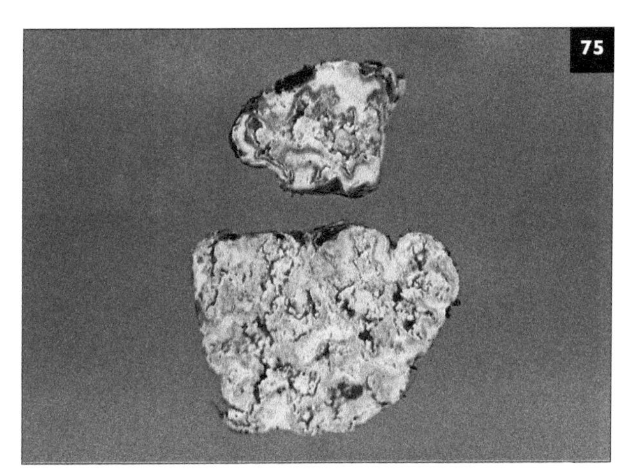

75 A false truffle, the ascocarp of *Diehliomyces microsporus*, cut through showing the corrugations which account for one of the common names for this disease, 'calve's brains'.

76 False truffles of the fungus *Diehliomyces microsporus*, forming predominantly between the casing layer and the compost.

anywhere within the casing or compost, including on the surface of the casing (77). They form on the outsides of spawned blocks, or bags of compost, next to plastic covers. The ascocarps contain many small ascospores about 5 microns in diameter and following their breakdown by bacteria the ascospores are released into the casing or compost. Spread of the ascospores is most likely to occur in drainage water, and with debris.

The optimum temperature for the germination of ascospores is 30°C, although they will germinate at much lower temperatures (16°C). Spore germination is said to be stimulated by the presence of actively growing mushroom mycelium. A short initial period of 1–2 days after spawning at 30°C is enough to stimulate the development of False truffle. Such conditions can occur in hot weather, especially with bag or block compost, or in any crop where cooling is not adequate. Ascocarps do not form at 16°C, but compost is rarely this cold, with the possible exception of bag and blocked compost which has cooled during transit, especially in cold winter conditions.

Ascospores were once thought to be very heat- and chemical-tolerant, but this is not the case. False truffle, in contaminated compost, is eliminated when heated to 60°C for 2 hours. Treatment at lower temperatures or for a shorter time is not effective. In this respect, False truffle should not survive phase II, and the general improvement in phase II may account for the decreased incidence of this disease. A high level of gaseous ammonia at kill (at least 450 ppm) is also important in the elimination of the fungus.

The most likely initial source of the pathogen is soil mixed with compost ingredients. An ineffective phase II would then allow spores to survive. Once established on a farm, the ascospores may be on any surface, but are most likely to be on containers such as trays and shelves, as well as in the dust on the floor of the growing room. *Agaricus bitorquis* can be seriously affected by False truffle; the higher growing temperature used for this species is a significant factor.

Control

If False truffle is a persistent problem on a farm, the phase II process should be carefully examined, and temperatures checked throughout the compost. Poor phase II together with soil-contamination of phase I compost (often by water draining from nearby land), are major factors in the occurrence of the disease. The phase I area should have a smooth concrete surface to allow thorough power washing between batches of compost. Soil contaminated straw is also a possible source. Such contamination occurs as a result of storage on a soil surface, splash during periods of heavy rain before harvest, or harvesting with the roots of the plants attached.

Strict attention to hygiene, an effective phase II, and the use of filters (2 micron) in both the phase II spawning and spawn-running areas, will prevent False truffle occurring. A satisfactory level of gaseous

77 Ascocarps of *Diehliomyces microsporus* on the casing surface.

ammonia (450 ppm), measured 3 hours after the maximum kill temperature, is also very important. Covering the spawn-run with paper and regularly spraying it with Formalin (0.25% *see* p. 56) will minimize the risk of late contamination. High compost temperatures for periods of 24 hours or more during spawn-running must be avoided, as these will encourage the germination of ascospores.

All affected crops must be cooked-out or chemically treated. A temperature of 65°C maintained for 12 hours is adequate to kill ascospores, but it is important to make sure that this temperature is achieved throughout the compost. In addition, trays and wood used to make shelves must be thoroughly cleaned by heat or with a disinfectant at the end of cropping. Lining trays with polythene, which is replaced for every new crop, can minimize the risk of carry-over. If heat is not available, Formalin fumigation of empty trays can be effective. Thorough net cleaning in shelf systems is also essential. Special attention should be given to disinfecting floors. There are no satisfactory fungicidal treatments of compost or casing either before or after the disease has occurred.

All *A. bisporus* spawns can be affected.

Action points

- Correct phase II temperatures must be achieved.
- Aim for at least 450 ppm of free ammonia measured 3 hours into phase II.
- Prevent soil contamination of compost by ensuring that the compost yard is adequately cleaned between batches of compost, and that the straw is free from soil.
- Use absolute filters during phase II, at spawning, and during spawn-running.
- Avoid high temperatures of 30°C and above during spawn-running.
- Achieve an effective cook-out.
- Cook-out empty trays and shelves.
- Remove all debris from trays and shelves and chemically treat between every crop if cook-out is not available.
- Line trays with polythene.
- Thoroughly disinfect the floor of affected cropping houses if they have not been heat-treated.
- Do not leave newly spawned bags and blocks stacked for longer than is absolutely necessary, especially in summer, in order to minimize the risk of the compost overheating.

Trichoderma diseases

A number of different species and strains of *Trichoderma* producing a range of symptoms are found in mushroom culture. Some are very damaging and others much less so. The relationship between *Trichoderma* spp. and the mushroom is not fully understood and almost certainly varies with the species and strain. Some are pathogenic, certainly on mushrooms, and others are probably pathogenic on mushroom mycelium. The genus is well known for its mycoparasitic ability, and for the production of toxins and antibiotics. Although the relationship between the mushroom and those species and strains associated with compost colonization is far from clear, for the sake of simplicity, they are considered to be pathogens.

The species most frequently associated with the mushroom crop are *Trichoderma atroviride, T. aureoviride, T. hamatum, T. harzianum, T. inhamatum, T. koningii, T. longibrachiatum, T. pseudokoningii, T. viride,* and *T. virens*. Strains of each of these species may differ in their ability to affect mushrooms. Important strains of *T. harzianum* have been recognized by various researchers, and two of these strains, designated *Th-2* and *Th-4*, are associated with compost colonization. Other strains of *T. harzianum* as well as other species may on occasions colonize compost. It has recently been proposed that the strain *Th-2* be renamed *T. aggressivum* f. *europeaeum*, and *Th-4* be renamed *T. aggressivum* f. *aggressivum*. The new names for the types that affect compost are used in this book.

Symptoms

On casing, wood, debris, and mushrooms

Trichoderma species are characterized by the production of very large quantities of spores that vary in shades of green (hence the common name Green mould), and it is these that are seen during mushroom production. Green mould frequently occurs on the wood of shelves and trays, particularly after heat treatment. New wood, which is still relatively high in stored carbohydrate, is particularly likely to be affected. The mould growth begins as a prolific development of white mycelium, which within 2–4 days turns green as the spores are produced (78).

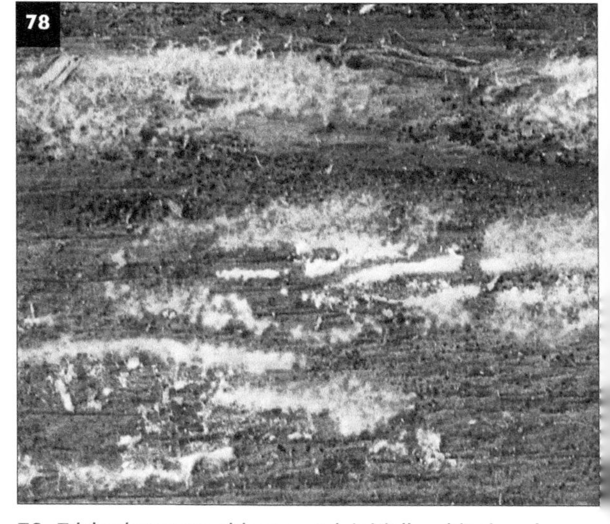

78 *Trichoderma* mould on wood, initially white in colour, but turning green as the spores are produced. A number of species can grow well, especially if the wood has not been seasoned or treated with preservative.

Many *Trichoderma* species have been found on wood or on debris, and include *T. atroviride, T. viride, T. virens, T. longibrachiatum, T. harzianum, T.*

koningii, *T. hamatum*, and *T. pseudokoningii*. Extensive colonization of the wood of trays often leads to cap spotting of mushrooms, particularly those developing nearest to the inoculum. Spots that are pale brown to grey in colour, without a clearly defined edge, have been associated with *T. koningii* and *T. pseudokoningii* (**79**). These spots are generally small, not more than 5 mm diameter, and often numerous. Others can be larger, dark brown and with a diffuse edge (**80**). These symptoms have often been associated with *T. harzianum*, but other species may be involved. Severely affected mushrooms sometimes show a dry decay originating from this type of spot (**81**). The cap spot symptom can be confused with Cobweb spotting. It is not known whether all species of *Trichoderma* cause spotting.

A similar range of species has been found on stalks of mushrooms left on beds as well as on mushroom debris on the casing surface. Green mould on trash including stalks and debris is common, and is not significant, except as a source of spores for cap spotting.

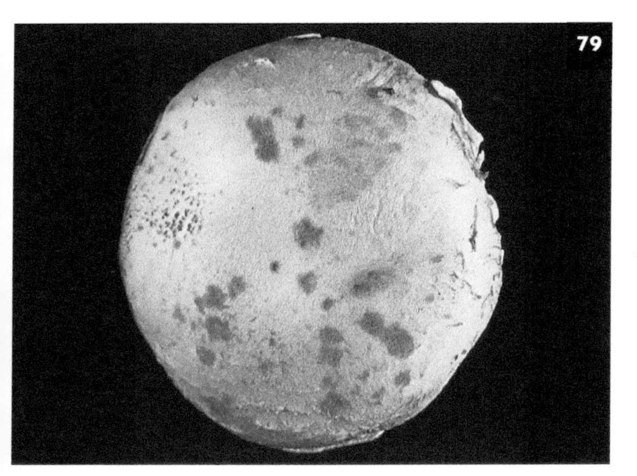

79 Pale brown spots on a mushroom (*Agaricus bisporus*), caused by *Trichoderma koningii*.

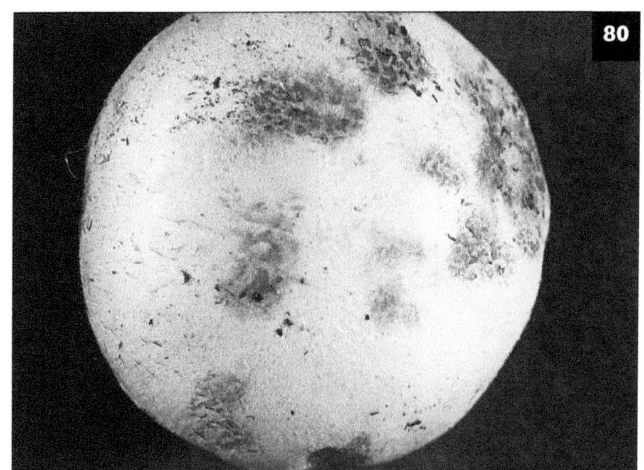

80 Large dark brown spots on a mushroom (*Agaricus bisporus*), caused by *Trichoderma aggressivum*.

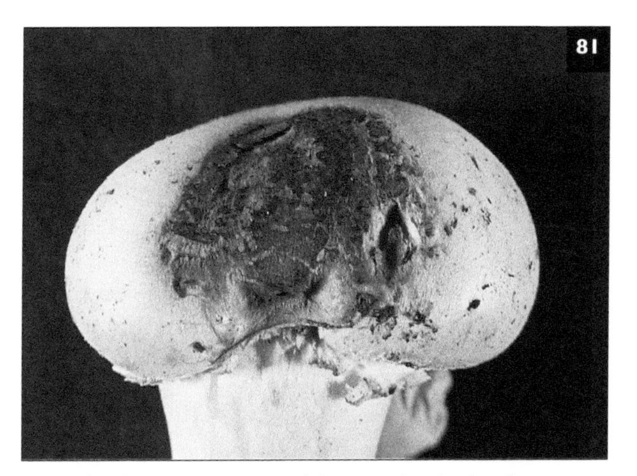

81 A dry decay on *Agaricus bisporus* developing from a spot caused by *Trichoderma aggressivum*.

82 *Trichoderma aggressivum* colonizing compost. Note the small tufts of white mycelium and the large green sporing mass of the fungus.

83 *Trichoderma aggressivum* sporing prolifically on the surface of the casing of a bag crop.

84 Severe attack of cap spotting associated with Trichoderma compost mould (*Trichoderma aggressivum*).

85 A very heavy infestation of Red pepper mites associated with Trichoderma compost mould.

Symptoms of compost Trichoderma

The most damaging form of *Trichoderma* is that associated with the colonization of compost. The name Trichoderma compost mould distinguishes this disease from all others caused by *Trichoderma* spp. Green mould, a name sometimes used to describe the compost problem, is confusing as it is also used to describe the more general symptoms listed above. The species largely concerned with Compost mould is now known as *T. aggressivum* (syn. *T. harzianum*). The disease is often first seen at the spawn-running stage, and appears as a dense white mass of mycelium, sometimes on the top of the compost, or between the base boards of trays, or through the sides of bags or blocks or through the netting on shelves (**82**). This mould also appears on the surface of casing, generally by the first flush stage of the crop. It turns dark green as spores are formed (**83**).

Trichoderma compost mould causes considerable reduction in yield, and also quality may be affected as a result of cap spotting (**84**). Frequently associated with outbreaks of this disease are large populations of red pepper mites (*see* p. 159), which feed on the spores of the *Trichoderma* spp. (**85**). The presence of red pepper mites may be the first indication of Trichoderma compost mould.

The pathogens and disease development

The pathogens

Trichoderma spp. that can cause cap spots are commonly found in many different habitats, including soil and on organic material. They produce chains of sticky conidia. Spores can be readily spread by air movement mainly on debris, and by insects and mites particularly red pepper mites, by personnel, and on all the various containers and equipment used on a mushroom farm. The growth requirements of various species are not the same, and the optimum growth temperature varies from 22°C to 28°C. Some *Trichoderma* spp. grow particularly well at pH below 6, especially if the nitrogen level is low. A carbon to nitrogen ratio of 22 or 23:1 is said to favour the growth of moulds in compost including *Trichoderma* spp.

T. aggressivum, associated with Trichoderma compost mould, has not been commonly found outside of mushroom farms. It is characterized by its fast growth rate (more than 1 mm per hour at 28°C on agar), by a longer vegetative phase than other species, and a higher optimum temperature for growth (28°C). In Europe, the strain most frequently found has been called *T. aggressivum* f. *europeaeum* (*Th-2*), whereas in North America the same symptoms are caused by the closely related *T. aggressivum* f. *aggressivum* (*Th-4*). These symptoms are not confined to these species, and in Australia both *T. inhamatum* and *T. hamatum* cause Trichoderma compost mould.

It appears that various strains and species of *Trichoderma* are capable of colonizing compost to a greater or lesser extent. Their exact identification does not affect the control measures that are applied. Precise identification is however useful in the study of the epidemiology of the disease, especially where it might be possible to trace the source of the inoculum.

Disease development

Cap spotting and the colonization of debris and wood usually occur with prolonged periods of wetness. New wood with a high carbohydrate status is particularly vulnerable. Mushroom spots develop when there is an abundance of inoculum and the mushrooms remain wet for periods of at least 2 hours. Water splashing spores from affected wood to mushrooms is a common means of transfer. Leaving mushrooms to over-mature on the beds also encourages the development of Trichoderma cap spots. Mushrooms with Trichoderma cap spots are often found at bed ends, or between the compost and the sides of the container, especially when the compost has shrunk away from the sides.

Compost mould begins when phase II compost is contaminated, or becomes contaminated at spawning. *Trichoderma* spp. grow well on carbohydrate, and the grain of the spawn is therefore an important food source and very vulnerable. It has been shown that as few as one *Trichoderma* contaminated spawn grain per tray can initiate an epidemic. Once in the compost, the mould is able to colonize large areas, and may rely on the even distribution of the spawn grains for its food source. Systems of culture using bags or blocks wrapped in polythene appear to be particularly vulnerable. This may be because of the condensation of water on the inside of the polythene, which not only provides an ideal environment for spore germination, but also helps to distribute the spores. In addition, bag and block culture involves transport from the composter to the grower, and the compost temperature may not be controlled during transit. A period of high temperature near to the optimum for *Trichoderma* (28°C) increases the chances of the mould developing. There is some evidence that high carbohydrate supplements added at the time of spawning may also encourage the development of Trichoderma compost mould.

At low temperatures (15°C) and at high temperatures (30°C), *Trichoderma* is less disadvantaged than is mushroom mycelium. The first 24 hours after spawning are very important in such circumstances, as the growth of mushroom mycelium away from the grains is delayed, and the grain is unprotected and very vulnerable to *Trichoderma* colonization.

Once *Trichoderma* is established within the compost, high temperatures during spawn-running increases the rate of compost colonization. Fully run compost does not generally become affected by *T. aggressivum*.

Control

On debris, wood and mushrooms

Elimination of the sources of inoculum is difficult if not impossible. Specific measures for the control of *Trichoderma* on wood and on the surface of casing are generally not worthwhile, except when a persistent problem results in cap spotting. Debris, including old stalks and discarded mushrooms, should be removed from the cropping surface and care taken to retrieve mushrooms that have fallen between bags and down the edges of the compost at the sides of trays and shelves. Wood is generally most vulnerable to colonization when it is still relatively green. If it has had a preservative applied to it before use, this will help to minimize colonization. Cooking-out at the end of the crop usually removes surface contamination, but deeper-seated mycelium in wood may not be killed and will appear during spawn-run. Reducing the temperature of the cook-out from 70°C to 65°C is said to minimize the production of spores on the wood surface. It is possible that the lower temperature treatment allows the survival of some antagonistic organisms that inhibit the development of *Trichoderma*.

Tray-dips, containing azoconazole or propiconazole, are effective in reducing sporulation on wood. Sometimes a carbendazim spray to woodwork (label permitting) during the spawn-running period will help to reduce the growth of the fungus, but routine application of this fungicide should not be encouraged because of the likelihood of fungicide resistance developing.

The most effective way to control cap spotting is to avoid conditions of wetness and poor evaporation from the mushrooms as they develop. The same conditions that favour Bacterial blotch (*see* p. 96) also favour the development of fungus-induced cap spotting, and the same environmental control measures are equally effective (*see* p. 97).

Trichoderma compost mould

Control of compost mould can be difficult, particularly once it is well established on a farm with the accompanying red pepper mites that are able to carry the spores from one crop to the next. The movement of personnel, flies, and small animals, also assists spread. To achieve control, *Trichoderma* spores must not survive phase II, and must not contaminate the compost after cool-down: they must not enter the system during spawning, and above all, they must not come in contact with the spawn.

In order to minimize the risk of contamination during the early stages of production, every effort must be made to reduce the inoculum levels on the farm. This can only be done by the strictest attention to hygiene (*see* Chapter 3). Contamination of compost during the period from cool-down to the first week of spawn-running must be avoided. Spores are killed by normal peak heat temperatures, particularly if the level of gaseous ammonia during kill is at least 450 ppm and preferably more. Temperature treatment without ammonia may not be enough to kill all the spores. Once the compost has been effectively treated, filtered air must be used to regulate its temperature. The spores are not smaller than 2 microns in diameter, and are therefore removed by absolute filters.

It is crucial that the spawn is handled cleanly, and that the spawn-hopper is regularly surface-sterilized. Trays must be thoroughly cleaned before use, either by cooking-out at the end of the crop, or by combining this with chemical treatment (*see* p. 48). Treated trays can become recontaminated if left standing on the farm. Spawning lines, conveyors, and transportation equipment must be thoroughly surface-sterilized immediately before use. Lining trays with polythene, which is changed for every crop, may also help. Disinfection of all structures and concrete surfaces around the farm is essential.

Compost received in bags or blocks should not be allowed to get cold, or to overheat. It is important to maintain the temperature at about 25°C.

Clusters of red pepper mites should be covered with salt, or burnt off with a flame gun as soon as they appear, in order to prevent them carrying spores into clean areas. Pickers' overalls should be cleaned daily, either by machine washing or by heating them in a tumble drier on maximum setting for 30 minutes.

Formalin, phenolics, and quaternary ammonium compounds are reasonably effective in killing *Trichoderma* spores. Formalin fumigation (4 ml/m^3, *see* p. 56) in spawning areas and in cropping houses is one of the most effective ways of cleaning structures, but care must be taken to make certain that the structures do not leak and that conditions are optimal to make the most of this treatment. One very important point is to fumigate in dry conditions, and not immediately after washing down with water.

Because carbohydrate-rich spawn grains are very often the starting point for *Trichoderma* in compost, effective control can be achieved either by eliminating the grain source, or by treating the grain with fungicide. A grain-free spawn has been developed in the USA for this purpose (SpeedSpawn). The manufacturers claim considerable success in the control of Trichoderma compost mould with it.

Although fungicides are generally ineffective, spawn treatment with carbendazim (0.23 g of 50%

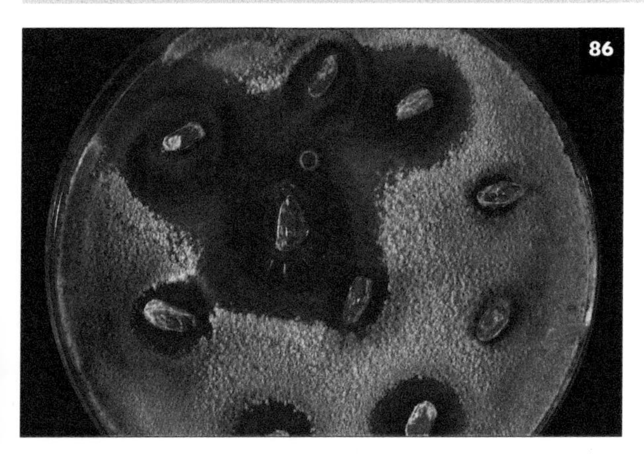

86 A bioassay for the coating of a benzimidazole fungicide on grain spawn. The fungicide treated spawn grains are placed on an agar plate previously seeded with a fungicide sensitive isolate of *Trichoderma* sp. The zones of inhibition show which grains have been treated and the size of the zone is proportional to the amount of fungicide on the grain.

a.i. carbendazim per kg of spawn) does give very good control. A similar recommendation for the use of a related material, thiophanate methyl (Senator, Topsin M), exists in Canada and some States of the USA. It is important to apply the fungicide evenly to the grain of the spawn. The efficacy of application can be checked in a laboratory (**86**). Evenness of application of carbendazim is improved if the fungicide is first diluted in 25 g of gypsum or hydrated lime per kg of spawn. A dedicated food mixer or something similar is required to evenly mix the fungicide with gypsum/lime. A cement mixer, suitably cleaned, is ideal for the application of the fungicide to spawn, but care must be taken to avoid physical damage. About six turns of the mixer is all that is needed providing the fungicide is added gradually. When treated, the spawn should be used immediately.

A cause for concern was reported from the USA in 2005: although apparently still sensitive to thiabendazole, strains of *T. aggressivum* var. *aggressivum* were found to be resistant to thiophanate methyl and benomyl. Recently, imazalil (Fungaflor) has been shown to be a satisfactory alternative fungicide for the treatment of spawn but, so far, this fungicide has not been registered for use on the crop. Great care must be taken in the use of any fungicide to control Trichoderma compost mould as the chances of resistance developing are increased with greater use. In this respect it is unwise to routinely treat supplement with any fungicide.

By far the most effective treatment of an affected crop at termination is by cook-out. If this cannot be done *in situ*, then great care must be taken when moving the crop to the cook-out room (*see* p. 47). Where heat is not available, tray crops must be well sprayed-off before they are emptied. Thorough cleaning of the farm after emptying is essential. The same applies with bag and block crops although it may be possible to tie bags before they are removed from the cropping area. Cropping houses must be very well cleaned between crops.

Action points

On casing, wood, debris, and mushrooms

- Treat new timber with a registered wood preservative before making it into trays.
- Reduce the temperature of cook-out to 65°C.
- Do not leave mushroom debris on beds or on the floor.
- Do not allow mushrooms to remain wet for periods of 2 hours or longer.
- Remove mushrooms that have fallen between bags or discarded mushrooms.

In compost

- Measure gaseous ammonia 3 hours into phase II (kill) and make sure the level is always 450 ppm or more.
- Filter air into phase II rooms, spawning areas, and spawn-running rooms.
- Disinfect the spawning line before use.
- Disinfect the spawn hoppers and make sure that spawn is handled with clean hands or clean gloves.
- Treat spawn with carbendazim or thiophanate methyl according to local regulations.
- Do not use compost supplements with added benzimidazole fungicides.
- Do not allow spawned compost to overheat/get cold.
- Cover spawn-runs with paper.
- Cover patches of red pepper mites with salt.
- Cook-out affected crops and disinfect cropping houses.

Aphanocladium cap spotting

Brown spots on the cap have been associated with the fungus *Aphanocladium album*. Other species (*A. aranearum* var. *sinense* and *A. dimorphum*) have been linked with similar symptoms in China.

Cap spots are light to dark brown in colour, roughly circular, up to 10 mm in diameter, and sometimes sunken (87, 88). The edge of the spot is usually diffuse but can be dense, and the spots may be very similar in appearance to those caused by *Cladobotryum* or *Trichoderma* and, in some circumstances, Bacterial blotch.

When a crop is severely affected, a large proportion of mushrooms show symptoms. Spots most frequently occur following conditions of high humidity and low evaporation, and often increase in frequency when open mushrooms are allowed to develop. The same conditions favour the development of other spots and it is not unusual to find Verticillium or Bacterial blotch in a crop affected by Aphanocladium. Sometimes in high humidities, white aerial mycelium of *Aphanocladium* grows on the affected tissues. Some observers have reported small pieces of organic material (a possible energy source) in the centres of spots, although these may not be essential for spot development.

The exact relationship between *Aphanocladium* spp. and the mushroom is not fully understood. It may be a weak pathogen, but various researchers have reported difficulty in reproducing symptoms with this fungus. Spots can be induced when concentrated spore suspensions are placed on the caps of developing mushrooms. The fungus is recorded as a pathogen of other fungi, as well as of some insects, and it has also been found in compost and in chicken manure. The frequent association of *Aphanocladium* with *Verticillium* and other fungal pathogens of mushrooms, as well as with Bacterial blotch, may indicate a positive association between *Aphanocladium* and these other pathogens.

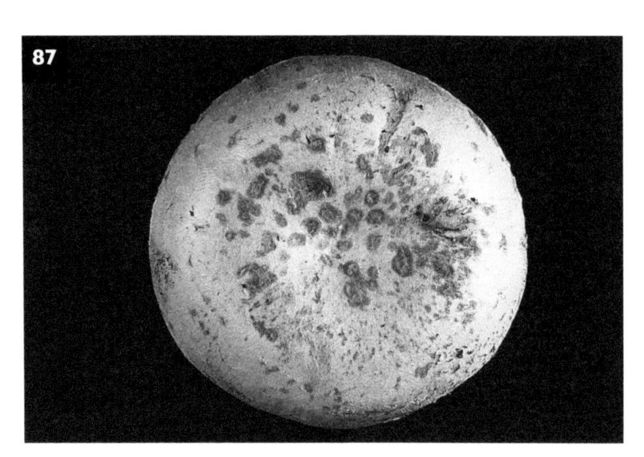

87 Aphanocladium cap spotting. The spots can be pale brown in colour, fairly small and sometimes numerous. The centre of the spot is sometimes sunken.

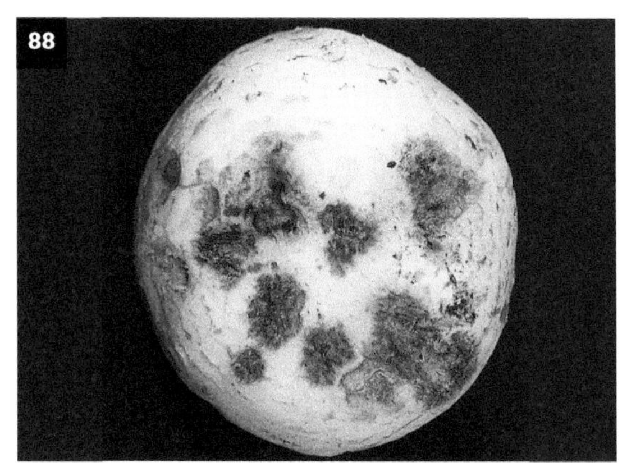

88 Aphanocladium cap spotting. These spots are dark brown and large. There is also an associated amount of casing debris which might be a factor in their development

Control

This disease is occasionally serious but can largely be controlled by increasing the evaporation rate from the crop (*see* p. 96, control of Bacterial blotch). Fungicides recommended for the control of dry bubble disease (*see* p. 75) are likely to give control, although there is evidence that some isolates of *Aphanocladium* are resistant to thiabendazole and prochloraz manganese while being sensitive to carbendazim.

Other diseases

The following diseases are, at present, of minor importance. But their incidence may change and the following description may help their recognition.

Gill mildew

Stunting of the gills with a growth of white mycelium on their surfaces is said to be the characteristic symptom of this disease. It has been suggested that a fungus (*Verticillium lamellicola* syn. *Cephalosporium lamellicola*) is the cause but there is some doubt about this, and symptoms have never been reproduced experimentally.

89 Peeling back of the stalk, discolouration of the cap and mycelium colonizing the mushroom are all symptoms of Shaggy stipe (*Mortierella bainieri*).

Shaggy stipe

This disease is caused by the fungus *Mortierella bainieri*. *Mortierella reticulata* has also been reported from mushroom crops, but has been associated with surface mycelial growth on spawned compost, and is thought to be a weed mould rather than a pathogen.

The most characteristic symptom of Shaggy stipe is the peeling of the stalk of the affected mushrooms, giving a shaggy appearance (89). The stalk and cap are usually discoloured, becoming dark brown as the disease progresses. The coarse grey-white mycelium of the pathogen can usually be seen growing over the affected mushroom tissue, and also over the surrounding casing (90). It is superficially similar to

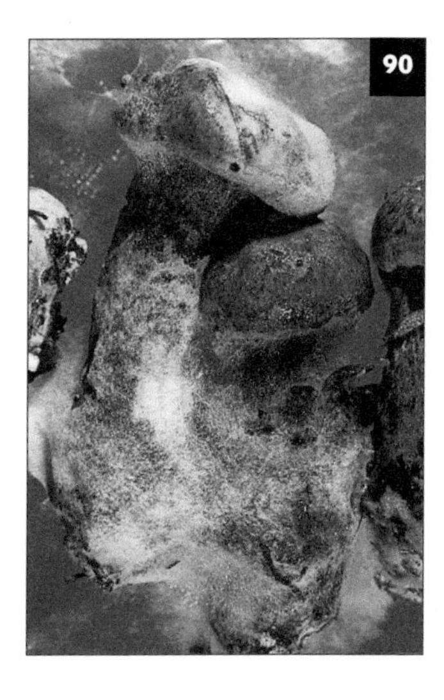

90 Coarse grey mycelium is very apparent with Shaggy stipe disease (*Mortierella bainieri*).

Cobweb (*see* p. 76), but is distinguishable from it by the colour of the coarse mycelium and the symptoms shown by the affected mushrooms. The cap may also develop a brown blotch, often surrounded by a yellow ring.

M. bainieri produces masses of sporangia and sporangiospores, and it is these that spread the pathogen. It is likely that the spores are both air-borne and water-borne. The fungus is a common soil dweller, and soil is likely to be the main initial source.

M. reticulata has been found in spent compost, which is suspected as being the main source of this fungus in the USA.

The most severe outbreaks of Shaggy stipe occurred during the period when benzimidazole fungicides were used extensively.

All diseased mushrooms should be covered in salt (*see* p. 42). No further treatment should be necessary. Generally very little yield loss occurs.

Hormiactis cap spot

There have been a number of reports of the fungus *Hormiactis alba* causing spotting in mushroom crops, but the disease is very uncommon. The symptoms consist of irregular brown spots or blotches about 10 mm in diameter, occurring anywhere on the cap surface. Hormiactis cap spot is probably the same disease as that previously attributed to a species of *Ramularia*. It is readily controlled by treatment with a benzimidazole fungicide, or disease removal and the use of salt.

Gliocladium diseases

Symptoms caused by *G. deliquescens* and *G. virens* have been reported in Europe and India respectively.

G. virens has been shown to have close relationships with some species of *Trichoderma* and for this reason has been renamed *Trichoderma virens*.

G. deliquescens is said to start on the casing layer where it usually grows from undeveloped pins. White fluffy mycelium spreads quickly over the surface, eventually forming a bright green (*Penicillium*-like) luxuriant weft as the spores are formed.

While developing mushrooms may be overwhelmed by the spreading mycelium, initial infection of older mushrooms occurs on the base of stalks, which have a characteristic black colour and may crack. The fungus may completely colonize the mushroom from the initial basal infection; a greyish white mycelium is then seen on the outside. Less frequently, mature mushrooms may show dark brown blotch symptoms. This spotting is said to resemble Aphanocladium spotting and mild blotch.

G. virens forms brown necrotic lesions which may penetrate from the margin of the cap to its junction with the stalk. The stalk may then be colonized and shows necrosis and splitting.

Should control be necessary, these pathogens are sensitive to the benzimidazole fungicides.

Conidiobolus disease

A prolific growth of very fine white mycelium has been seen on the casing on a few occasions accompanied by a significant decrease in yield. The fungus involved, *Conidiobolus coronatus*, is a known pathogen of insects and man, but was first recorded on cultivated mushrooms. There is some doubt about it being a pathogen of mushrooms, although in one experiment where it was added to casing a considerable reduction in yield occurred. There are some reports of other *Conidiobolus* species being pathogenic to agarics.

Large populations of *Conidiobolus* can occur on farms where fly populations are very high. When many dead flies fall onto the casing, for instance near to electrostatic fly traps, they appear to be a good medium for the growth of the fungus, and may be its source.

This mould may be more significant than is generally realized, and could account in part for the reduction in yield associated with very large phorid fly populations.

CHAPTER 5

Bacterial Diseases

Introduction

Bacteria play a very important part, both good and bad, in mushroom production. Many types of bacteria are present in compost, and many of them play an important role in all stages of compost production; they are also involved in the initiation of mushrooms. *Pseudomonas putida* has been shown to be important in mushroom initiation and is present in high concentrations in casing. Other bacteria in the casing have been implicated in the rapid degradation of fungicides.

There are a number of well known diseases of mushrooms caused by bacteria, in particular Bacterial blotch.

Blotch, Bacterial blotch or Brown blotch

Blotch is one of the most widespread diseases of *Agaricus bisporus*, and occurs in every country where the crop is grown. It also affects *A. bitorquis*. Yield losses are the result of its affect on quality, both before and after harvest. It can be particularly severe and difficult to control in countries such as Australia, with high summer temperatures. Generally it is now much less of a problem in Northern Europe. Its decline may be, at least partially, due to the widespread use of effective air conditioning.

Blotch is generally caused by the bacterium *Pseudomonas tolaasii*, which is universally present on mushroom farms. Recent research in New Zealand and in Canada suggests that other bacteria may be able to produce similar symptoms. Various species of bacteria have been implicated. In Canada, isolates of *Serratia liquefaciens*, *Cedecea davisiae,* and one isolate of *Pseudomonas putida* were found to be pathogenic. It seems likely that a variety of bacteria are capable of producing yellow to dark brown discolouration of the mushroom cap, but the most common is *Ps. tolaasii*.

The most important factor governing the development of Blotch is persistent wetness on the surface of mushrooms. For this reason, the disease is always worst where air conditioning is poor, and particularly in climates where it is more difficult to maintain good evaporation.

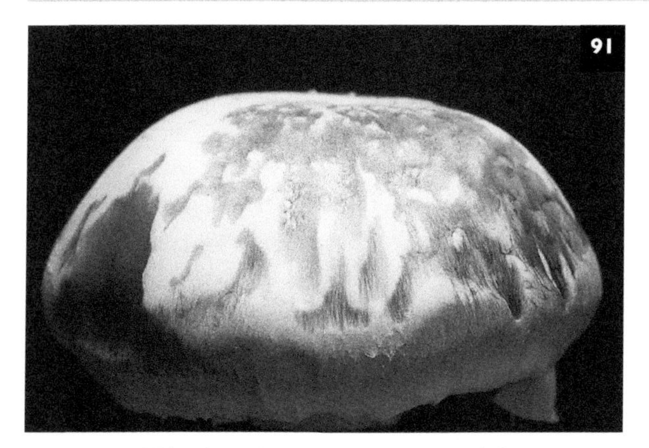

91 Bacterial blotch with variation in colour of the spots from light brown to very dark brown.

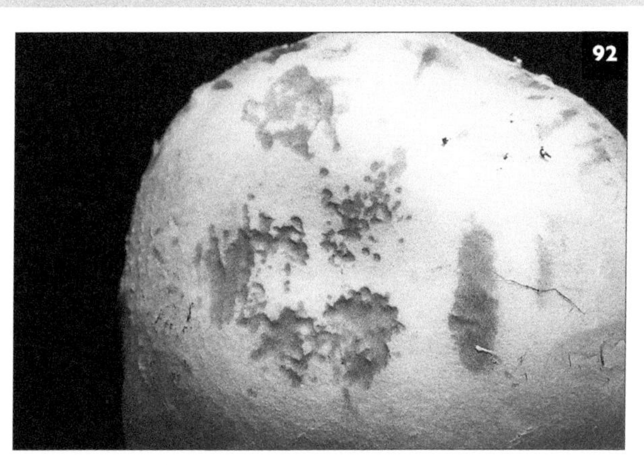

92 Bacterial blotch with large and small spots on the cap surface.

93 Bacterial blotch, a severely affected crop with spots of all sizes and colours.

94 Bacterial blotch with the smaller affected cap splitting as it dries.

Symptoms

As the name suggests, the most characteristic symptom is the occurrence of brown blotches on the surface of the cap. These vary in colour from light brown (**91**) to dark brown or almost black. The whole of the cap may be affected, or more frequently there are discrete spots of various sizes and numbers (**92** and **93**). Sometimes they are small (1–2 mm) and very numerous. Affected mushrooms are slimy to the touch. If drying conditions occur after the Blotch symptoms have developed, the affected mushrooms distort and the caps split (**94**). The stalk may also be affected. The distribution of blotches on the cap often coincides with the parts of the cap which remain wet the longest, i.e. on the margins or in gullies where two mushrooms are touching.

Disease development

The pathogen appears to be endemic, probably surviving between crops on surfaces, in debris, on tools, on flies, and on structures. It may also be introduced into each crop in the casing, as it is a natural inhabitant of peat, chalk and possibly sugar beet lime. *Ps. tolaasii* colonizes mushroom debris such as discarded mushrooms, stalk bases, and dead pins on the bed surface. Such sources of the pathogen can make a major contribution to disease incidence if conditions for infection are favourable. The pathogen can be dispersed on debris, on mushroom spores, by flies, by mites, on the hands of pickers, on boxes, and on pickers' equipment. In addition, once the disease has become established in a crop, watering will disperse the bacteria very effectively. Whatever the source, Blotch develops on every farm when the growing conditions favour its development.

The pathogen population increases from the time casing is applied to the crop. *Ps. tolaasii* is able to utilize organic gases produced by the growing mycelium as its carbon source, and in this way the population builds up during case-run. The more vigorous the spawn-run the quicker the population increases. Very rapid propagation of the pathogen occurs on the mushroom surface, even when the mushrooms are very small. The most important period for infection starts when the developing mushrooms are pin-head size (2.5 mm diameter). A critical population of 10^6 bacteria per gram of casing is necessary for infection of these pin-heads. Microscopic infection points are formed within a few hours, especially at temperatures of 20°C or above. The cap surface is a weave of mycelial threads, and within this mesh water can be held. Small amounts of water that are not visible, or enough to make the mushroom feel wet, are sufficient for bacterial propagation followed by infection. However, symptoms do not appear unless water persists on the cap surface. A wet cap follows watering or may result from fluctuating air temperature and short periods of dew-point conditions. Microscopic lesions develop rapidly if water is present for more than 2 hours, and spots may become visible within a few more hours. Conditions which favour Blotch occur frequently where environmental control and air circulation are poor and in climates and at times of the year when the ambient atmosphere is saturated.

Disease development is a two-stage process: first, infection which results after the population of the pathogen on the cap surface has reached a critical concentration; second, subsequent spread from the microscopic infection points, which is dependent upon surface wetness.

Little can be done to influence the first of these two stages, and it is likely that many apparently healthy mushrooms are microscopically infected at harvest but show no obvious symptoms. The control of Blotch is essentially the prevention of the second stage, that of lesion development from the initial infection points. This is done by environmental control aimed at preventing mushrooms being wet for more than 2 hours. It is a tribute to the growing skill of most growers that this is consistently achieved, so that Bacterial blotch is not often a serious problem.

If mushrooms are developing Blotch symptoms, but the environmental conditions for Blotch development become less favourable, lesion spread is arrested. In some circumstances, particularly when symptom development is stopped totally, a black pit on the mushroom surface is all that is visible. This cause of the pit symptom is probably the most common, and the isolation of other organisms from mushrooms with pits probably results from secondary colonization of damaged mushroom tissue.

If infected, but symptomless, mushrooms are harvested and stored in conditions favourable for disease, such as high ambient temperatures and humidities, browning will develop. Thus mushrooms that are apparently healthy at harvest can show marked symptoms a few days later (*see* p. 108, Post-harvest browning).

Control

As the bacterium appears to be universally present on all farms, it is impossible to control the disease by the elimination of the source. It is likely that casing is always a primary source.

To guard against extra contamination of the casing, the site chosen to store the ingredients and mixtures should be covered and protected from wind-borne debris (*see* p. 117). Where the disease is a persistent problem, heat-treating the casing may be beneficial, but should be considered to be a last resort treatment.

By far the most important factor in disease control is the minimization of surface wetness. It is important to get good evaporation and drying at all times, but particularly when water is applied to

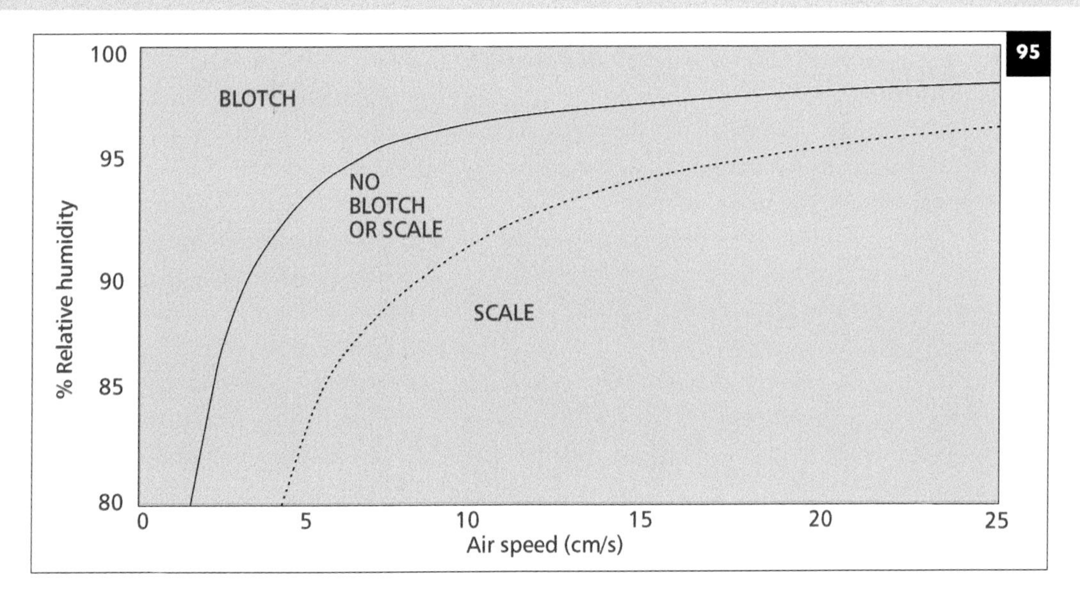

95 The relationship between air speed, relative humidity, Bacterial blotch and mushroom scaling. After G.E. Bowman, 1991.

developing mushrooms. Surface evaporation depends upon the air temperature, the amount of water in the air, and the speed of air movement over the surface of the crop. There is a delicate balance between the relative or absolute humidity, and the speed of air-flow over the crop. If air movement is too slow, evaporation is reduced and conditions suitable for Blotch development may occur. In contrast, when air movement is too fast, mushrooms become scaly. The relationship between these factors is summarized in (**95**). From this diagram it can be seen that when the relative humidity is 80%, a very slow air speed (4 cm/s) is all that is needed to avoid Blotch, and at this humidity it requires air movement of only 5 cm/s to cause scale. In contrast, at 95% RH, Blotch will develop at air speeds up to 6 cm/s, and it requires a speed in excess of 15 cm/s to cause scale. At 95% relative humidity, an ideal air speed is 10 cm/s or 1 m/10 s.

Some growers control the rate of evaporation using the absolute humidity of the air rather than the relative humidity (*see* p. 58). The success of this system is dependent upon very accurate temperature control. An upper limit for the amount of water in a kilogram of air is set, and it should never be allowed to exceed this, irrespective of the chosen temperature. The level of about 11 g of water per kilogram of air at normal cropping temperatures allows an air deficit of between 2 and 3 g of water which is generally all that is needed for enough evaporation to avoid Blotch.

Whether control is by relative or absolute humidity, it is essential to have efficient environmental control equipment, because small fluctuations in temperature can result in condensation, followed by disease development.

Monitoring evaporation from the surface of developing mushrooms is not easy, although various attempts have been made to devise appropriate equipment. A Piché evaporimeter has been found by some growers to provide a useful guide. With this simple apparatus, it has been shown that blotch is rarely a problem if the daily rate of evaporation between the rows of shelves, boxes or bags is at least 0.8 ml of water (*see* **47**).

Watering the casing with a chlorine solution is used to keep the bacterial population down with the aim of preventing it reaching the critical level for infection. To stand a chance of being successful, such a treatment must start early and not be left until the last watering before pinning or, worse still, until disease symptoms appear.

Various preparations of sodium or calcium hypochlorite have been commonly used. Sodium hypochlorite products usually contain 10–12% available chlorine. When hypochlorite is added to water, it dissociates into hypochlorous acid and chlorine. It is the hypochlorous acid which is most bactericidal. The dissociation is pH-dependent. At low pH (below 7) hypochlorous acid predominates and at high pH it is mainly chlorine. An ideal pH is 6.5 and this must be achieved after the hypochlorite has been added, as this raises the pH. If the pH is too high, the water can be acidified by the addition of hydrochloric or citric acids. If there is a strong smell of chlorine at the time of watering, it is likely that the pH is too high.

As the chlorine content of the concentrate decreases in light and at ambient temperatures, it is important to store the main container in the dark and in a cold store. The available chlorine content of the concentrate can be checked (*see* Appendix 2). Chlorine is used at 150 ppm applied at every watering (*Table 5*), and this concentration must not be exceeded. In the UK, but this may not be the same in all countries, there is a one-day harvest interval for the application of chlorine water at 150 ppm to developing mushrooms.

In addition to sodium and calcium hypochlorite, chlorine dioxide is used in some countries and has been shown to be effective, but is not approved for use on mushrooms everywhere.

Although some are effective, no antibiotics have been registered for use on the crop.

A product originally known as Conquer but later called Victus, has been successfully used for the biological control of Bacterial blotch. It is a bacterium (*Pseudomonas reactans*) which is antagonistic to the Blotch organism. The same organism is used in the white line test to identify *Ps. tolaasii* (*see* p. 32, and 28). It is applied to the compost just before spawning, and also to the casing at the onset of pinning, and after picking the first flush. This product is not available or registered in all countries.

There have been recent experiments on the control of Blotch with bacteriophage. Phages are viruses of bacteria, and when applied to the casing are able to limit the population of the bacterium. So far no commercial products are available.

None of the commercial strains of *A. bisporus* is resistant to Bacterial blotch, although there is some variation in the susceptibility of *A. bisporus* isolates in collections made from wild populations. These may, at some time in the future, form the basis for resistant strains.

TABLE 5 Obtaining 150 ppm available chlorine using sodium hypochlorite concentrate

Per cent available chlorine in the concentrate	Quantity in ml to be added to 100 litres
15	100
14	107
13	115
12	125
11	136
10	150
9	167
8	188
7	214
6	250
5	300
4	375
3	500
2	750
1	1500

Action points

- Adjust the conditions within the cropping house so that evaporation is taking place from the surface of the developing mushrooms.
- Avoid water remaining on the surface of mushrooms for more than 2 hours. In such conditions Blotch will develop almost irrespective of how thoroughly other aspects of control are applied.
- Make sure that temperature control is as precise as possible, as stable temperatures will prevent dew-point being reached.
- Apply 150 ppm chlorine with every watering.
- Remove diseased mushrooms and debris, avoiding spread of the bacterium by using the correct procedures (*see* p. 41).
- Store all casing materials, before and after mixing, in an area free from contamination.

Ginger blotch

This disease is caused by *Pseudomonas gingeri*, a bacterium closely related *to Ps. tolaasii*. The ginger colour of the blotches, which start as a yellow-brown discolouration and do not change with age, distinguishes this disease from Brown blotch (**96**). Affected mushrooms are sticky and slimy to the touch. It is unusual for both Blotch diseases to occur in the same crop. As far as is known, the ecology of the organism, means of spread, and conditions which favour disease development are similar to those for *Ps. tolaasii*. Casing materials are probably the most important primary source.

Control measures are identical to those recommended for Bacterial blotch.

96 Ginger blotch the red-brown symptom on affected mushrooms.

Mummy disease

Mummy disease was common, and some farms had a persistent problem resulting in considerable reductions in yield. Now it is very infrequently seen. There is still doubt about the precise cause of this disease, and it is practically impossible to confirm by laboratory test. Both American and Dutch work has indicated that bacteria are involved in Mummy disease. The most frequent is a species of *Pseudomonas* closely related to *Ps. tolaasii*. However, in experiments *Pseudomonas aeruginosa* has also been shown to produce similar symptoms, and large populations of this bacterium have been reported in the compost of Mummy diseased crops. Researchers have also speculated that fastidious-like bacteria (*see* p. 34) could play a role, as it has been shown that bacteria-like structures, which have proved to be very difficult to culture, can be found within affected mycelium. Experiments to prove the pathogenicity of various bacterial isolates, including those above, have either failed or been only partially successful. It seems likely that there is a relationship between the bacterial population in the compost and the development of Mummy disease, but so far the exact nature of this relationship has not been clearly demonstrated. An imbalance caused by cultural or other factors could lead to large bacterial populations that are capable of causing Mummy disease. Compost wetness is one such factor.

Symptoms of Mummy disease are very similar to those of the virus diseases of the crop, and recent diagnostic techniques used in the identification of viruses may throw new light on the cause of Mummy disease.

The way the disease has been reproduced most consistently is by mixing compost or casing from an affected crop, with clean compost, at the time of spawn-running. There is no evidence of spread in any other way.

Symptoms

The most characteristic feature of the disease is its fast progression along a bed, which is quoted as being 10–25 cm of bed length per day. When crops are grown in trays, bags, or blocks, spread is less obvious, although this may not be the case where there is mycelial contact between the units.

Affected mushrooms die and become very dry, with the internal tissue discoloured, often with brown streaks. When cut transversely, the affected stalks sometimes show small (1–3 mm) dark brown spots. The caps may be distorted and are commonly tilted (97). At the base of the stalk, the mycelium is very stringy, and there is often a basal swelling together with a growth of fluffy white mycelium. When affected mushrooms are removed from the bed a large amount of casing adheres to the base of the stalk. It is said that pickers are able to detect affected mushrooms by the feel of the stalks when cut, probably because of their dryness. Additional symptoms recently described include the appearance

97 Mummy disease with brown streaks in the dry stem of the affected mushroom and a clump of mycelium attached to the base.

of a brown/red layer close to the base of the stalk within minutes of it being cut, as well as a delay in pinning and harvesting. These latter symptoms, together with the general appearance of the affected mushrooms, are very similar to those associated with virus diseases. Also, affected crops, like those with virus diseases, do not recover (*see* p. 170, False or Crypto-mummy disease).

The only distinguishing feature of Mummy disease is its rate of spread, but observations on this aspect of the disease date back many years when mushrooms were grown in beds. Mummy disease is more difficult to diagnose in modern growing systems, except where shelves are used. Positive confirmation is made more difficult by the absence of a laboratory test.

Original reports of Mummy disease indicated that mushroom cropping stopped once symptoms were seen, but this is not always the case, and some affected crops continue to yield, although poorly, for the duration of the crop.

Traditionally, it was believed that wet or anaerobic compost favoured the development of Mummy disease. Top trays in phase II, which become wet as a result of condensation dripping from the roof, appear to be particularly vulnerable.

Because of the difficulty of consistently reproducing disease symptoms, it has not been possible to do systematic tests on spawn susceptibility. Any apparent differences between strains may reflect their inability to anastomose, and therefore transmit the causal agent. Until reliable tests are developed, reported differences between strains must be treated with caution.

Control

As soon as Mummy disease is suspected, attempts should be made to localize it. Traditionally this was done by digging a trench across the bed on either side of the affected area about 1.5 m in advance of the symptoms. The trenches were 20 cm wide, and all the compost from the trenches was removed from the crop or placed on the surface of the affected area. The isolated area of bed and the exposed bases of the trenches were drenched with 0.5% Formalin, and the surface of the affected area covered with salt or lime.

With a tray or bag crop, the affected bags or trays can be isolated or removed, making sure that there is no mycelial contact with adjacent healthy bags or trays. Alternatively, the surfaces of affected bags or trays can be totally covered with salt.

At the end of the crop, the house should be thoroughly cooked-out. Trays and woodwork should be heat-treated or treated with disinfectants before reuse. If only a few trays are affected, it is worth marking these and giving them a prolonged dip treatment in disinfectant, particularly if heat-treatment is not possible.

Action points

- Isolate the affected areas by digging trenches.
- Treat the affected areas with 0.5% Formalin (*see* p. 56).
- Cook-out at the end of cropping, preferably using steam (*see* pp. 46–48).
- Mark affected trays and pay special attention to their disinfection as well as areas of bed where symptoms have been seen. It is essential to kill all mushroom mycelium (*see* p. 48).
- Make sure that the compost is not excessively wet during phase II.

Pit

This disease is sporadic in occurrence and rarely causes large crop losses. There may be a number of causes. *Ps. tolaasii* is associated with pitting in certain conditions (see p. 96), and there are reports of other bacteria, including *Bacillus polymixa* and *Erwinia caratovora,* being involved. Some workers have implicated mites and nematodes. It seems most likely that Pit is primarily a bacterial disease, and is the result of infection at an early stage in the development of the mushroom. If the development of the decay is arrested, and the mushroom continues to grow, a dark sunken area or a pit results. The disease is often associated with conditions of very poor evaporation followed by a period of good evaporation. The presence of pits reduces the quality of the crop, often making it unsaleable.

Symptoms

Generally Pit does not appear until fairly late in the crop, and is most frequently seen in the third or later flushes. Very small (1 mm or less), dark (often black) slimy pits, or needle-like pricks, occur on the cap surface of an otherwise healthy mushroom. They may increase in size with the growth of the affected mushroom. The pits may penetrate the cap to a depth of several millimetres. Pits are randomly distributed on surfaces of caps, and vary in number from one to ten or more per mushroom (98). It is likely that the symptom is initiated by infection at a very early stage in the development of the mushroom.

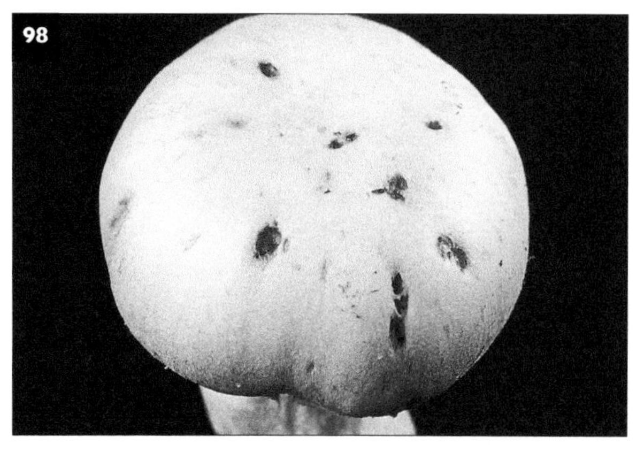

98 Bacterial pit. The pits varying in size and number.

Control

As this disease is thought to be caused by a bacterium, those measures described for Bacterial blotch (*see* p. 96) are usually used for its control. Frequently the disease is not sufficiently serious to justify any action, and it is often transitory in appearance.

Action points

- As for the control of Bacterial blotch (*see* p. 96).
- Check phase II temperatures to make sure that these are high enough to kill pests such as mites or nematodes.

Drippy gill

This disease is caused by the bacterium *Pseudomonas agarici* and is very uncommon. Like the previous bacterial diseases, it seems to be associated with poor evaporation.

Symptoms

The gills are attacked by the bacterium even before the veil of the mushroom is broken. When gills are infected at an early stage, they are much paler in colour than normal. Small dark brown or black spots appear on the side and bottom edges of the gills, eventually increasing to 2 mm in diameter or more. At the centre of each spot is a creamy-grey droplet which may increase in size to bridge the gap between the gills (**99**). These droplets contain large concentrations of bacteria. In severe attacks the droplets may coalesce on and between gills to form ribbons of slime with progressive collapse of the gill tissue (*see also* cecid damage, p. 151). Sometimes early infection is associated with the inhibition of mushroom development and consequent distortion. The stalks of affected mushrooms develop longitudinal splits which may be more than 2 cm long. Their inner surfaces are shiny, and microscopic examination reveals the presence of bacteria. As affected mushrooms mature the splits turn dark brown.

The presence of bacterial ooze on the gills of unopened mushrooms, and also the bacteria in the splits in the stalks, suggest that the pathogen is probably systemic.

Little is known about the conditions that favour this disease. However, it is possible that flies, pickers, and water splash spread the organism within the crop. Because the bacteria may be systemic, infection may take place at a very early stage in mushroom development, or the bacterium may originate from infected mycelium.

Control

Little if anything is known about the disease, and there are no specific control measures. Those recommended for Bacterial blotch may be useful (*see* p. 96).

99 Drippy gill with droplets of bacterial ooze on the gills and also spanning the gap between the gills.

Bacterial soft rot

A very rapid soft rot of *Agaricus bisporus* has been shown to be caused by the bacterium *Janthinobacterium agaricidamnosum*. A similar rapid soft rot of *Agaricus bitorquis* is caused by *Burkholderia gladioli* pv. *agaricola*. Both are very rare, but occur locally and sporadically causing disastrous losses, especially after harvesting. Both pathogens are easy to isolate from mushrooms showing the first signs of the disease, but as the decay progresses, other bacteria also are found.

Symptoms

The first signs of disease on *A. bisporus* are depressed areas on the surface of mushrooms. These rapidly turn into brown oozing pits which eventually consume the whole of the mushroom, sometimes overnight. Affected mushrooms may be scattered in the crop or single mushrooms may occur (**100–102**). Similar symptoms have been seen in crops of *A. bitorquis*.

In recent Belgian experimental work, the bacterium was introduced into the compost at spawn-run, into the casing after casing application, and onto the casing at pinning. Symptoms occurred irrespective of the time of inoculation. The most severe disease resulted when infection took place during pinning. Infection of the pins resulted in a total collapse of mushrooms within 48 hours. Infection of nearly mature mushrooms also resulted in symptoms.

Studies of the spread of the pathogen by air currents, by watering, and by contact showed that the bacterium was easily spread by water splash over a distance of 54 cm, and also by contact, but not by air currents. Contamination of hands resulted in transfer up to 40 times. Mechanical harvesting has been associated with some of the most severe outbreaks.

Sciarid flies, although vectors of bacteria and fungi, are thought to be insignificant as a means of spread of this pathogen as they are not attracted to decaying mushrooms.

Control

The use of chlorine has been shown to be ineffective; Formalin treatment has not controlled this disease. However, improving evaporation is effective.

Action point

- Improve evaporation.

100 Bacterial soft rot affected mushrooms in advanced stages of decay. Photograph by kind permission of John Burden.

101 Bacterial soft rot with the colour variations that can occur. Photograph by kind permission of John Burden.

102 Bacterial soft rot: affected mushrooms change from healthy to completely decayed in a very short time, the decay often starting from a sunken lesion. Photograph by kind permission of John Burden.

Internal stalk necrosis

This disease has been reported more frequently in recent years as growers strive to improve market quality. It occurs at about the 1% level in many crops, but occasionally at harvest a significant proportion of mushrooms show symptoms of internal stalk rot (103). The bacterium *Ewingella americana* has been consistently isolated from affected mushrooms, and the symptoms have been reproduced by injecting developing healthy mushrooms with this organism. *E. americana* is usually associated with sewage and has also been found in shellfish (molluscs). Although it is not known to be an ingredient of casing or compost, detailed studies of these materials in relation to this organism have not been done.

Symptoms

Brown discolouration of the central tissue of the stalk is seen when affected mushrooms are cut (**104**). Examined in longitudinal section, the brown tissue extends from the base of the stalk to the cap, and rarely penetrates into the cap tissue. Affected mushrooms may be wet in appearance, but frequently, at harvest, the brown tissue is dry and has completely collapsed leaving a hollow centre (*see* hollow stems on p. 170 and sciarid damage on p. 143).

This disease is sometimes associated with water-logging of the stalks of mushrooms at an early stage in their development, although by the time the symptoms are seen the stalks may no longer be water-logged. Abnormally low compost temperatures after pinning the first flush may be a contributory cause. Unlike Soft rot, the decay does not normally extend beyond the central tissue of the stalk. It is possible that the same conditions give rise to both Soft rot and Stalk necrosis, and the symptoms depend upon which organism colonizes mushroom tissue.

Control

Because there appears to be an association between the crop's water relations and the development of Internal stalk rot symptoms, it is important to maintain good evaporation from the bed surface at all times (*see* p. 58). The application of chlorine at 150 ppm may help to minimize the bacterial population.

Action points

- Improve evaporation.
- *See* Bacterial blotch, p. 96, for other relevant points.

103 Internal stalk rot (necrosis). This symptom is not seen until the mushroom has been harvested.

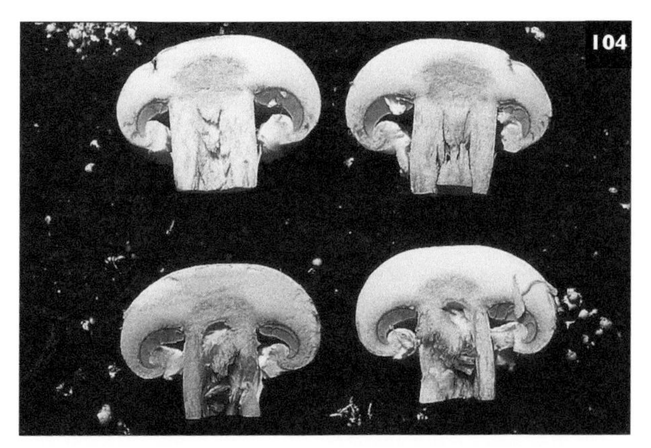

104 Internal stalk rot (necrosis): the rot of the central tissue of the stalk often accompanied by hollowness.

Post-harvest browning

When the whole mushroom cap discolours after harvest, it is not strictly a disease symptom but the first stages of senescence (105). However, some of the bacteria known to cause disease are implicated in its development. Most of the organisms associated with Post-harvest browning are related to *Pseudomonas fluorescens*. *Ps. tolaasii*, the cause of Bacterial blotch, is in the same group as this bacterium. There are reports of considerable reductions in incidence resulting from the application of bactericidal products to the growing crop.

Symptoms

As the name implies, affected mushrooms develop off-white to brown colours after harvest. In some instances, this may appear overnight, while in others it may take more than one day. The discolouration results from the bacterial population that is present on the mushrooms at the time of harvest. If the harvested mushrooms are stored so that their surfaces remain wet, especially if they are not cooled quickly, discolouration develops. The severity and speed of development of the symptoms will depend upon the initial bacterial population as well as the conditions of storage. It follows that any treatment which minimizes the surface bacterial population at the time the mushrooms are harvested will reduce the risk of discolouration developing.

105 Bacterial browning of mushrooms before harvest. Affected mushrooms feel slimy and are generally wet. This type of bacterial decay is generally a secondary effect of another disease, e.g. Virus or Mummy.

Control

In many countries there are no specific measures recommended for the control of this problem. However it is likely that the use of chlorine, as for the control Bacterial blotch, will help to minimize bacterial discolouration.

In some countries the application of chlorine dioxide and calcium chloride has been shown to improve post-harvest whiteness and therefore quality.

Action point

- Use chlorine dioxide and/or calcium chloride in countries where this is permitted.

CHAPTER 6
Viral Diseases

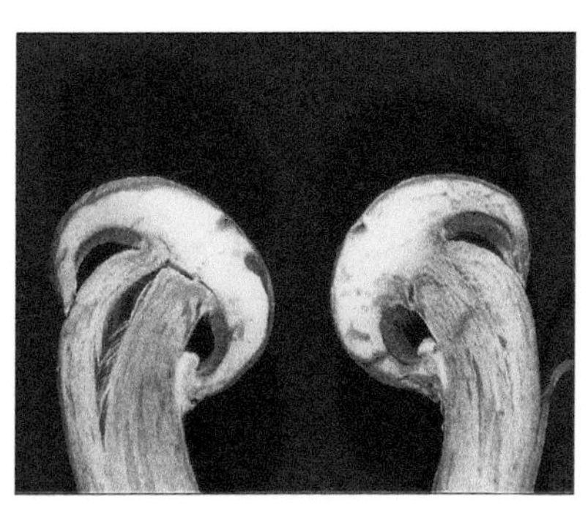

Introduction

There is still some uncertainty about mushroom virus diseases and the absolute proof that the virus particles seen within the mushroom mycelium, and within the sporophores, are entirely responsible for the symptoms attributed to them has still to be proven. Purified virus preparations have not been introduced into the mushroom and the symptoms reproduced (Koch's postulates – isolation, inoculation, symptom production, reisolation – have not been satisfied). The situation has become somewhat more complex since the discovery of Mushroom virus X disease, because virus particles have not been found, although the genetic material (ribonucleic acid) is present. There is still much to learn about virus diseases of the mushroom. Had current technology been available years ago when Mummy disease and False mummy disease were relatively common, a virus component may have been found in these. They certainly show many of the same characteristics of the virus diseases.

This chapter is concerned with the two diseases of *Agaricus bisporus* attributed to viruses and which, for the sake of simplicity, we refer to as virus diseases. Much of the epidemiology and control of them is common to both. More specific points are described under each disease.

Names of mushroom virus diseases

For many years growers have universally referred to 'virus', in spite of early investigators using various names such as Die-back, Watery stipe, and Brown disease. A now out-of-date numbering system was used to name the mushroom virus particles in sequence as they were discovered (*MV1* 25 nm, *MV2* 29 nm, *MV3* 50 X 19 nm, *MV4* 35 nm, and *MV5* 50 nm).

Research has shown that 35–36 nm spherical virus particles (one nanometre, nm, is one millionth of a millimetre or 10^{-9} m, and the slight size differences reported probably reflect the difficulty in measuring accurately), are responsible for La France disease. This virus has been named *La France isometric virus* (LIV), the name referring to the farm in the USA where it was first found, and the shape of the virus particles. Only one other particle type has been given a name and that is the 50×19 nm particle which has been called *Mushroom bacilliform virus* (MBV). As far as is known it is not involved in symptom production.

A 'new' disease has, for the moment, been called Mushroom virus X disease (MVXD). It has a symptom syndrome similar to La France disease. No virus-like particles are associated with Mushroom virus X disease, and the viral ribonucleic acid (RNA) found in affected crops is not the same as that found in particles of *LIV* (*see* Chapter 2). Similar naked dsRNA (double stranded) viruses have recently been found in a number of green plants and fungi. The name Endornavirus has been proposed for this newly recognized family of viruses. The relationship, if any, between *LIV* and Mushroom virus X disease is as yet unknown. In Britain, the appearance of MVXD has been coincidental with the absence of *LIV*, implying that there may be some inverse link between the two. MVXD has been found in a number of European countries.

The pathogen *Mycogone perniciosa* is also known to have a virus, although it has not been shown to be related in any way to either of the two mushroom virus diseases.

Epidemiology of mushroom virus diseases

Finding only one virus type in a mushroom is rare, and many apparently healthy crops have small concentrations of a number. The significance of these small concentrations is not understood. There is no observable gradual build-up from low concentration to high followed by disease development, but when higher concentrations are found, disease often follows. Resultant action, usually in the form of improved hygiene, has been linked with a decline in the particle count and no loss in yield. Whether or not the two are positively linked has still to be proven.

Observations suggest that large numbers of air-borne mushroom spores or fragments of mushroom mycelium are consistent precursors of virus disease development. It is known that mushroom spores carry *LIV* and also some of the other types (25 nm spheres, MBV and probably MVXD). Two possible means of virus transfer are known: from mycelium to mycelium through hyphal fusion (anastomosis) with the transfer of the cell contents, and transfer in mushroom spores which germinate and pass virus to healthy mycelium, again by anastomosis. There is very good circumstantial evidence that both of these means of transfer are important in the development of virus disease on farms.

Anastomosis between mushroom strains is essential for virus transfer and may be commonplace or infrequent, depending upon the relationship of the strains. Those strains most closely related, e.g. mid-range hybrids, will anastomose. Experimental work indicates that off-white strains do not readily anastomose with smooth-whites. Anastomosis between different species of *Agaricus* is unlikely to occur.

Fragments of viable mycelium left on, or in, the woodwork of trays or shelves and on netting could anastomose with new mycelium, and transfer virus to the new crop. Fragments less than one cell in size will not survive, and it may be necessary to have mycelial pieces of at least several cells for survival and virus transfer. Such fragments will be considerably larger than mushroom spores and are unlikely to pass through filtration equipment, but may be present in areas where bulk run compost is handled.

The transfer of fragments of mycelium from one batch of phase III compost to another is also thought to be a significant factor in the development of virus disease. In the process of bulk spawn-running, and at any time when spawn-run compost is moved, small fragments of mushroom mycelium can become air-borne, and may then be disseminated in much the same way as mushroom spores. Handling phase III compost either on the farm or between farms is therefore a high-risk operation which can ultimately result in virus disease if all the necessary precautions are not taken. Similarly, modern cultivation techniques such as ruffling and cacing can also result in the transfer of fragments of mycelium and with them the risk of disease development.

Mushroom spores measure 5×7 microns and are produced in very large numbers in mushroom crops. It has been estimated that a single mushroom with an 8 cm diameter cap will produce 1,300 million spores. The numbers of spores in the air of a mushroom house has been measured at 10,000 spores per m^3, and even in areas where there is filtration to prevent their entry, spore counts of up to 10 per m^3 have been reported. Such spores are presumed to have escaped filtration by leaving cropping houses through open doors at harvesting etc. Calculations show that as few as 10 spores per m^3 can result in 10,000–100,000 landing on each tray during spawn-run. In experiments, as few as l0–100 *LIV* viruliferous spores per tray induced a virus problem.

Mushroom spores are stimulated to germinate by nearby mushroom mycelium. Some other fungi, such as *Peziza ostracoderma* (cinnamon mould), have been shown to exert the same stimulation. It is clearly very important to be able to eliminate mushroom spores as far as is possible. When dry, they can remain viable for long periods, and are difficult to kill by heat. In longevity studies, germination of mushroom spores has been recorded after 30 years of storage. *LIV*-carrying spores germinate more quickly and more abundantly than healthy spores: in longevity experiments they have remained viable for at least 14 years.

There is conflicting evidence on the thermal death-point of mushroom spores, varying from a complete kill at 54°C for 10 minutes to 65°C for 16 hours.

Dust on all mushroom farms will contain mushroom mycelial fragments, bacteria, mushroom spores, and spores of fungal pathogens. Dust is therefore a very important source of pathogens including viruses.

As far as is known, there are no insects, mites, nematodes, or fungal vectors of mushroom viruses, but flies and other insects can distribute virus-carrying mushroom spores. There are no known wild mushroom hosts of mushroom viruses.

Man is not directly implicated in the transfer of virus on the farm although the way the farm is built and managed can have a very significant effect.

La France disease

La France disease has been called Watery stipe, Die back, Brown disease, and X disease. The original common names of La France disease are infrequently used and the disease has generally been referred to as 'virus'.

Symptoms

Various symptoms have been recorded, but rarely are they all present at the same time. Distortions of the mushrooms may occur, including elongation of the stalks (drum-sticks), tilting of the caps, very thick stalks particularly at the base, and very small caps on normal-sized stalks (**106, 107**). Commonly mushrooms show brown discolouration of the internal tissue in the form of streaks (**108**). They may be watery in texture; hence one of the common names, but may dry on the bed and become spongy (**109**).

Sometimes the gills of affected mushrooms develop poorly and are light in colour. This symptom is very like that of Hard gill (*see* p. 172), while some of the other symptoms are like those of Mummy disease, and False mummy (*see* pp. 100, 170).

Affected crops may show a patchy appearance with poor mycelial growth into the casing. In contrast, growth in the casing may be normal but the pins fail to develop. With time, the mycelium in such patches becomes stringy, the compost is wet, and the mycelium degenerates. Weed moulds may then be found in such patches. Early maturity (premature opening) of mushrooms or slight discolouration may occur. The discoloured mushrooms are often grey to pale brown in colour and may be slimy to the touch. Some affected crops show no very obvious mushroom or mycelial symptoms, but suffer yield losses. When crops are affected at an early stage, yield losses are very considerable. Later infection reduces crop yield less, but affected crops may still be a source of the disease for other crops on the farm.

Since the hybrid strains have been commonly used, distortion symptoms are less frequent. La France disease has not been found in the closely related species *Agaricus bitorquis*.

Identification

Generally, a sample for testing consists of 10 mushrooms taken at random from a second or third flush. The dsRNA of virus origin is extracted, and a PAGE or modified PAGE test done to identify the RNA present (*see* Chapter 2). In this way a positive confirmation of *LIV* can be made. Electron microscopy may also be used to confirm the presence of virus particles. RT-PCR can be used to identify specific viral material, and such tests can be done on mycelium in phase III compost, thereby checking quality standards before the product is released. In future, it is likely that all mushroom virus testing will be done using RT-PCR or systems evolved from it.

Disease epidemiology

The initial or primary source of La France disease is unknown but the occurrence and distribution of the disease do not suggest a spawn source. However, spawn cannot be ruled out until the mechanism of disease development is fully understood. *LIV* is known to be transmitted in mushroom spores and mycelium. These are very important means of dissemination. One significant feature in the epidemiology of La France disease is a close association between the occurrence of severe disease and open mushrooms on the farm. Virus disease symptoms have frequently been seen about 6 weeks after large numbers of open mushrooms were present on the farm.

Wild mushrooms and other fungi or bacteria have not, so far, been shown to be sources of La France disease.

106 La France disease affected crop in the early stages of disease development. Note the distortion of the caps and streaking on the stems.

107 La France disease, a severely affected crop. Note the mushrooms are severely discoloured, some caps are tilting, other show small caps on long stalks (drumstick symptom).

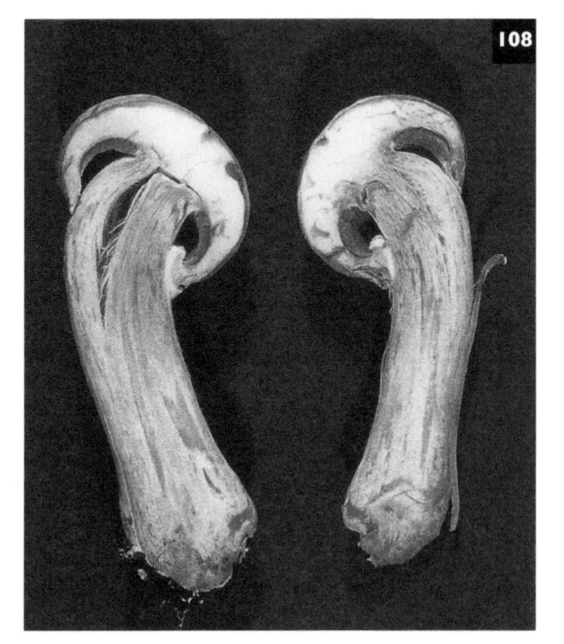

108 La France disease, internal browning and streaking within the stem of an affected mushroom.

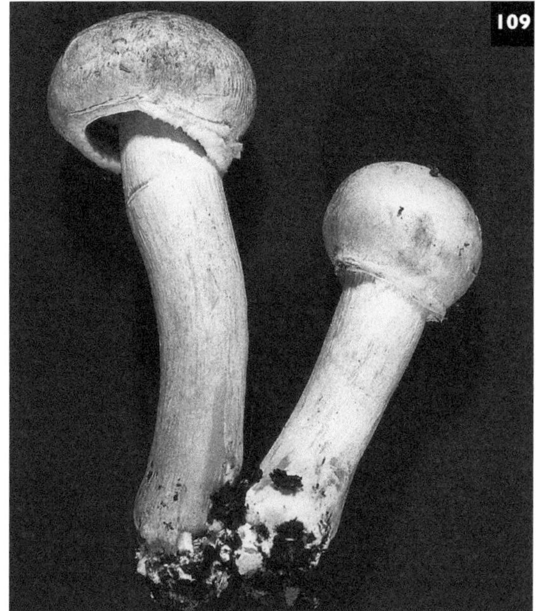

109 La France disease, watery symptom on the stalk of a distorted mushroom.

Control

The aim of a virus control programme is to ensure that no mushroom spores or viable mushroom mycelium comes in contact with the new crop. Spores and mycelial fragments are present in the dust on all mushroom farms. Dust as a source of inoculum cannot be over-emphasized. The introduction of dust at vulnerable stages of cropping will greatly increase the chances of a virus problem occurring.

At almost all stages in crop production, there are necessary procedures to minimize or eliminate the chance of La France disease being introduced into a new crop.

Phase I

Virus-infected spores from cropping houses, and dust deposited on compost ingredients or on phase I compost, may survive the composting process. Any spores that do survive should be killed in phase II. However, it is possible that some may survive phases I and II and in this respect it is important to have the phase I area away from the cropping houses, or fit appropriate exhaust filters on the cropping houses, to reduce the risk of spore contamination. The correct position of the yard in relation to wind direction, or the use of a windbreak, can also minimize or eliminate the risk of stack contamination. The chance of phase I being an ultimate source of virus is minimal, but cannot be totally discounted.

Phase II

It is likely that the combination of 60°C and an ammonia concentration of 450 ppm or more will be effective, but 60°C must be maintained throughout the compost for several hours in order to be certain that mushroom spores are killed (*see* p. 46). Mushroom spores are known to have survived traditional tray pasteurization. After conditioning and during the cool-down period, newly introduced spores may be able to withstand the compost temperature. It is therefore very important that any fresh air used in cool-down is filtered. Generally, absolute filtration is used (*see* p. 38), which removes all particles and spores of 2 microns diameter or larger. Very few, if any, mushroom spores should get through such filters. There is obviously little point in having expensive filtration if the phase II room is not air-tight. Holes in the ducting, gaps around the filtration unit, cracks in walls, badly fitting doors, can all contribute to the introduction of unfiltered air. To some extent, small cracks and holes in some parts of the structure are counteracted by the use of a positive air pressure within the room, so that air is always flowing out and never in, except through the filtration unit.

Spawning

This is a very high-risk stage for the introduction of virus as well as other pathogens. The area surrounding the spawning line should be thoroughly washed down, after each use of the line, by power washing with water, followed by disinfection and/or fumigation with Formalin (*see* p. 56). If the area is open, it is better to complete the washing down as near to the time of spawning as possible; ideally the night before, but this may preclude the use of Formalin. If spawning is done within a sealed building, washing down followed by Formalin fumigation after the surfaces are dry will give a very effective treatment. The building should be ventilated and kept under a positive pressure at all times during spawning. It is important that spawning equipment is thoroughly cleaned immediately before use and on the day of use, not the day before. This is best done by spraying it with a disinfectant. It is not necessary to use high volumes of water for disinfection, particularly if a thorough wash down with a pressure hose was done at the end of the previous use. A sprayer capable of delivering fine droplets will give all surfaces as good a cover as possible. A phenolic disinfectant, usually at 0.4% (*see* p. 54), or a QAC/glutaraldehyde mixture, are suitable for this operation. Spawn hoppers and supplement hoppers must be thoroughly cleaned, and spawn handled with clean hands and gloves. It is necessary to disinfect gloves at regular intervals. Workers on the spawning line should wear clean overalls.

Spawn-running

All the same conditions as apply to phase II are relevant here. Filtered air must be used, or air can be recirculated within an enclosed system with a cooler installed to regulate the air and compost temperatures. Where enclosed cooling systems are used, carbon dioxide levels in the rooms give a good indication of the integrity of the system. Low carbon dioxide levels indicate leakage is occurring and unfiltered air is entering the system.

Spawn-running rooms and ducting must be thoroughly disinfected and/or fumigated before use. During traditional spawn-running the surface of the compost can be covered with paper or polythene in order to prevent contamination by fungal spores, as well as preventing excessive water loss. Paper is often preferred to polythene because it absorbs moisture and prevents the surface of the compost becoming too wet. The paper should be lightly sprayed with 0.25% Formalin at 3- or 4-day intervals in order to kill spores landing on its surface. The last spray is best done the day before paper removal to prevent spores transferring to the compost surface.

Bulk spawn-running

Sites where compost is bulk spawn-run (phase III) are particularly vulnerable to virus disease. In order to minimize the risk of mycelial transfer, the emptying end of the bulk tunnels should be at the opposite end from the filling end, and in this way the spawning operation does not come in contact with the emptying operation. The building used for spawning phase II compost must be ventilated with filtered air, and kept under positive pressure at all times. This minimizes the risk of mushroom spores or fragments of mycelium entering the compost at spawning. Circumstantial evidence indicates that, where contamination occurs at this stage, a virus problem inevitably results. Over-pressurization of phase III emptying halls may be unwise as it can result in the distribution of mycelial fragments, which increase the risk of contamination of phase II compost at spawning. A neutral pressure may be necessary to prevent the exit of mycelial fragments and the ingress of mushroom spores. The recycling of mycelial fragments can be further minimized by thoroughly cleaning the phase III area after every tunnel is emptied. This can only be done effectively if the whole operation is enclosed. The structure can then be fumigated with Formalin using up to 4 ml for every m^3 of volume treated (*see* p. 56).

Phase III tunnels must be cooked-out at the normal cook-out temperatures. This will also clean the plenum which is often a source of compost debris. Cook-out is not necessary if the same tunnels are used to produce phase II compost, but, increasingly, phase III tunnels are solely dedicated to phase III production.

Casing

It is very important that the casing ingredients are free from contamination. Peat stored as wrapped bales or in enclosed bags, as well as bags of chalk, is likely to be 'clean', although the outsides of the bags and bales may be a source of contamination. Washing and disinfecting the exterior surface of the peat bags before they are opened will prevent debris on their surfaces entering the casing. Now that bulk-stored materials are more commonly used, these may be contaminated on site by the settling of air-borne mushroom spores, the spores of fungal pathogens or of air-borne dust containing these. It is therefore important to store casing materials in a separate area that is well protected from wind-borne contaminants. This area should be kept clean, with a foot disinfectant at its entrance, and should be power washed at regular intervals

During casing application modern shelf systems are vulnerable to contamination by air-borne dust containing spores or fragments of mycelium.

Post-casing and case-run

The crop is much less likely to suffer yield loss if infected at this stage, but two recent developments may be relevant.

First, it is now common practice to ruffle the casing when the mushroom mycelium has grown into it, in order to regulate yield and quality (*see* p. 14 and 10). This process is done either by machine or by hand raking. Mushroom mycelium is broken in this process and, inevitably, there is some movement of mycelial fragments along the bed. If small patches of virus are present, it is likely that affected mycelium will be spread. The chances of such spread can be minimized by cleaning the ruffling equipment at regular intervals. Ruffling machines should be washed with water at the end of each bed and sprayed with a disinfectant, or the paddles run in a trough of disinfectant. Hand-held rakes can be regularly dipped in disinfectant.

Second, the practice of adding mycelium to the

casing (casing inoculum and cacing – *see* p. 14) may also increase the risks. Obviously the added mycelium must be free from virus, but if the crop is already infected, it is possible that additional spread will occur during the process of incorporation. Cleaning of equipment, in a way similar to that already described for ruffling, should be practised.

Cropping

Infection of a mushroom crop after cropping begins is unlikely to result in symptoms or yield losses in that crop, but spores and mycelial fragments from such a crop can be a source of the virus for younger crops and in this way can be very important.

If the farm policy is to grow open mushrooms, or because of picking problems 'opens' have become unavoidable, mushroom spores will be produced in large numbers. An appropriate exhaust air filter capable of removing mushroom spores should be fitted to each house. Such filtration should be a routine if opens are continuously grown.

Crop termination

At the end of the crop, cooking-out *in situ* at 70°C for 12 hours will ensure that all mushroom spores and mycelium are killed. A shorter treatment time at this temperature is probably effective, but in order to be certain that all parts of the compost, the woodwork, and structure reaches a high enough temperature, it is generally necessary to treat for a 12-hour period. Other methods of crop termination are covered in Chapter 3 (*see* pp. 46–48).

Even after cooking-out it is advisable to treat empty trays with a further heat treatment as it only needs small fragments of viable mushroom mycelium or small numbers of spores to restart the disease. If cooking-out is not possible, trays must be cleaned by washing and chemically treated. After treatment, the trays should not be stored in an exposed position where they may become contaminated with mushroom spores and dust. If this is unavoidable, they should be sprayed with disinfectant again immediately before filling. Details of tray treatments are given in Chapter 3 (*see* p. 49).

Cropping practice

Except where the farm is growing open mushrooms, regular picking to minimize these is important. When opens do unavoidably occur, spore transfer into other crops must be prevented.

Washing down the farm at least once a week with a disinfectant ensures that mushroom spores on roadways are killed. Similar treatment should be given to vehicles and tools.

A regular schedule of filter and duct inspection is necessary. It is important to change both absolute and dust filters according to need, i.e. when they are dirty, not at predetermined intervals.

Recycling returnable plastic containers could be a potential problem if the containers are not cleaned before they are returned. Frequently, they are heavily contaminated with mushroom spores as well as those of fungal pathogens. Care must be taken to avoid storing such containers in vulnerable positions on the farm where spore transfer from them to crops is likely. They should have a high-temperature wash before they are returned to the farm (*see* p. 44).

Strains

A complete cropping cycle around the farm of off-white strains, or some smooth-white strains, or *A. bitorquis*, has, in the past, been recommended to reduce the general level of virus inoculum. However, these changes require such a major reappraisal of growing technology that they are unlikely to be worthwhile.

Action points

- Samples should be tested immediately a virus problem is suspected.
- Air used in the preparation of phase II and III compost must be filtered.
- Filters and ducts must be checked regularly for leaks. Filters should be checked and changed when necessary.
- Spawning area should be positively pressurized.
- All machinery used to handle phase II and phase III compost must be thoroughly washed and disinfected immediately before use.
- Equipment used for spawning must be cleaned immediately before use.
- Spawned compost should be covered before and after filling.
- Casing must be stored in a clean area in order to prevent dust contamination.
- Care must be taken to prevent contamination of the casing during its application.
- Cook-out of crops and structures must be effective (*see* pp. 46–48).
- Tunnels used for phase III compost production must be cooked-out to clean the netting, plenum, and structure.
- Blocking and bagging equipment must be thoroughly cleaned between uses in order to prevent mycelial carry-over.
- Buildings, roadways, and all concrete areas must be disinfected regularly (*see* pp. 53–56).
- Filters should be fitted to growing rooms to minimize the release of mushroom spores.
- Phase II and III winches should not be shared.
- Returnable plastic containers must be cleaned between uses.

Mushroom virus X disease

A disease characterized by non-productive patches of crop was first recognized in the UK in 1996–97, and has been called Virus X, Patch disease, and Patch syndrome. It has occurred on many farms, persisting on some in spite of the application of virus control measures. In this book we use the name Mushroom virus X disease (MVXD). This name indicates a degree of uncertainly about its cause, in particular whether one or more viruses, or even other causes, are involved in the production of the various symptoms. Since the initial outbreak, additional symptoms have been seen, including brown mushrooms in a white crop, early opening mushrooms, delayed cropping, distortions, and poor quality mushrooms. More specific identification, and a better knowledge of the cause of the various symptoms, may enable a more precise name or names to be given to the disease or diseases.

One distinct feature of MVXD is the very low concentration of dsRNA in affected mushrooms, compared with that of La France disease. So far there is no proof that any of the dsRNA bands are associated with a particular symptom although four bands have been consistently found in brown mushrooms. Some recent work indicates that these bands may have come from bacteriophage either in or on the affected mushrooms and not from the mushrooms. Further investigations are needed to resolve this complex situation.

Identification

Mushrooms from affected crops contain dsRNA, and up to 26 species (bands) have been found. These differ from the dsRNA bands associated with La France disease. No virus particles have been seen. For this and other reasons it is not clear whether the many dsRNA bands represent one or more viruses or are even incidental to the disease problem. However, because of the consistency of their occurrence, and their association with disease symptoms, it is presumed that they are significant, and their identification is therefore the first stage in the implementation of control measures.

Identification can only be done using a modified PAGE technique (*see* p. 34). The tests have proved to be slow and operator-sensitive. Samples of about 20 mushrooms are required (more than needed for a La France disease test) and even then the amount of dsRNA detected may be very small. PCRs are available for the identification of four of the bands. The numbers of bands in samples examined may vary considerably over a period of time, and from site to site, or even on the same site. In contrast, on some sites, the numbers of dsRNA bands found in affected crops are very consistent. Generally the greater the number of bands the more severe the problem.

One further characteristic of the disease which it shares with La France disease and Clusters (*see* p. 168) is a change in the type of growth, and reduced growth rate of mushroom mycelium in cultures made from affected patches (**110**). When anastomosed with healthy mycelium, the slow growth rate is transferred.

Symptoms

In some farms, virtually all of the symptoms occur together, although it is more common for one or two to predominate.

Patches

The symptom first noticed was patches of non-productive bed, not dissimilar to those seen with La France disease. These patches were circular or irregular, becoming more pronounced throughout the life of the crop. In shelves, the non-productive areas were sometimes in unusual patterns of swirls and strips along the shelf, with apparently normal mushrooms adjacent to the non-cropping areas (**111–113**). Non-cropping areas can be very extensive. In experiments, the patch symptom could not be reproduced using a mycelial culture taken

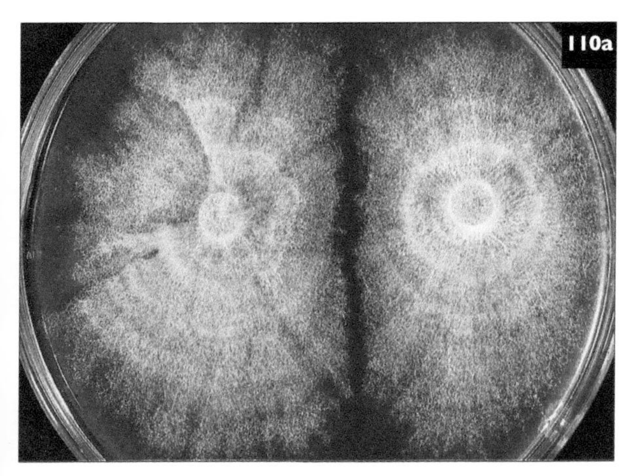

110 Growth of spawn on agar. (**a**) Healthy and (**b**) from a MVXD-affected crop. The cultures are the same age and the slower growth rate of the affected culture is also accompanied by a different growth form.

111 MVDX patches surrounded by apparently healthy mushrooms.

112 MVXD patch in a block crop: it is likely that this distinct patch was associated with an affected block.

113 MVXD with patches distributed in lines; they may also be in swirls.

from a crop with patch symptoms (*cf* brown caps). However, patches have occurred in crops when small quantities of affected compost were spread onto the surface of spawn-run compost.

Although mushrooms generally cease development in the patches, occasionally the patch is a delayed area of up to 4 days, not a non-productive one. The delay is sometimes so short that a 'hills and hollows' effect appears in an otherwise normal stand of mushrooms.

A disruption in the mechanism and development of pins could account for all of these symptoms, with the variation in degree and severity related to the concentration of the pathogen and/or the time of infection.

Early opening

Mushrooms with this symptom may appear to be quite healthy until harvest, at which time they open at a stage of development when they are normally tightly closed. Crops on some affected farms have consistently shown this symptom.

Associated with the symptom is an apparent near absence of a veil covering the gills. In some instances the edge of the cap appears frilled, and in others it is tightly rolled (**114**). Where the early opening symptom is found, it may affect a small number of mushrooms or almost the whole flush. It generally appears in the second flush, although it has been seen in all three flushes of an affected crop.

Crop delay

Crop delays of up to four or more days have been repeatedly associated with MVXD. This symptom may be the result of pinning disruption which, at the extreme, results in no mushrooms.

Brown mushrooms

Pale fawn to quite brown mushrooms occur in a white strain crop (**115**) which, in extreme cases, gives the appearance of white spawn contaminated with a cream or brown stain. The symptom usually affects the whole cap, but can occasionally discolour only a part of it. Affected mushrooms may be few in number or can be 40–80% of the total. The dsRNA pattern associated with this symptom has four distinct bands (18, 19, 21 and 22), as well as many others sometimes associated with other symptoms, adding weight to the suggestion that more than one pathogen is involved in the complex. The concentration of dsRNA appears to be greatest in the brown skin of the affected mushrooms. The tissue under the affected skin appears to be normal. The browning symptom has been reproduced in first and second flushes when healthy fully spawn-run compost was inoculated with a low level of infected spawn (0.01%) during bulk handling. Similar results were obtained when the casing was inoculated. Low levels of inoculum applied at spawning did not consistently reproduce the brown symptom, although some brown mushrooms appeared in the

114 MVXD with the edge of the mushroom cap tightly rolled.

115 MVXD-affected mushroom on the right with a healthy mushroom on the left. The overall pale brown discolouration of the cap is characteristic of some outbreaks, where it may be the only symptom.

second flush. Consistently low levels of brown mushrooms have occurred in mushroom crops inoculated late in the crop cycle. No brown mushrooms resulted when high inoculum levels (25% and above) were applied at spawning.

It is thought that environmental conditions, in particular evaporation, may affect the expression of this symptom.

Distortion and loss of yield

Many of the poor quality symptoms associated with La France disease have been seen with MVXD. An occasional symptom is the occurrence of a brown-red discolouration at the junction of the stalk and the cap. The colour increases in intensity with time after the cut has been made. A similar symptom has been recorded for mummy disease (*see* p. 100).

Disease epidemiology

Time of infection is important: early infection has the greatest effect, and infection after casing generally much less effect, with the exception of the brown symptom. Very small quantities of mycelial fragments are capable of causing this disease. Such fragments are generated when fully run compost is handled and these can be air-borne or become part of the 'dust' on the farm and, together with mushroom spores in the dust, constitute a very important source of the disease. Dust contamination at spawning, during spawn-run, and at casing can result in severe disease levels.

As with La France disease, Mushroom virus X disease is transferred by anastomosis between healthy mycelium and affected mycelium, or mycelium resulting from the germination of affected spores. In this respect mycelial fragments left on machinery can be a significant source of the disease.

Control

All the procedures listed for the control of La France disease apply. Any equipment used to handle both phase II and phase III compost must be meticulously disinfected between operations. It is better to have dedicated equipment for each operation, in particular winches.

Because of the very small amount of contamination needed to cause this disease, air-borne inoculum as mycelial fragments, mushroom spores or dust with both, is particularly significant.

Action point

- *See* La France disease, but with special attention to all those points which involve dust contamination.

Vesicle virus

Three dsRNA fragments without protein coats are widespread in hybrid and brown spawn strains, and appear to have no adverse effects on the growth of the crops. They are very frequently found in analysis of samples for MVXD and there are indications that all three are part of a single virus. They show no relationships with *La France isometric virus*.

CHAPTER 7

Moulds that Compete with Mushrooms

With contributions from Professor Albert Eicker, University of Pretoria, South Africa

- INTRODUCTION
- TRICHODERMA COMPOST MOULD
- PENICILLIUM MOULD
- SMOKY MOULD
- PYTHIUM – BLACK COMPOST
- OTHER MOULDS
- ACTION POINTS FOR CONTROL OF MUSHROOM MOULDS

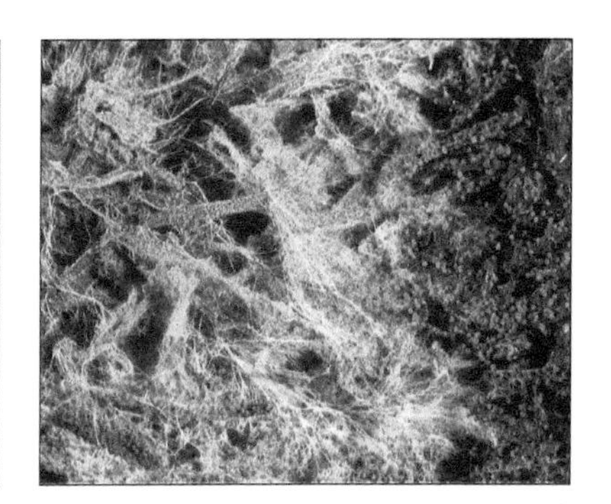

Introduction

The exact relationship between many moulds and the mushroom is not understood. For this reason the differentiation into fungi that are harmful and those that are not, is difficult. Before the era of bulk production of compost, moulds in mushroom crops were not uncommon. In the majority of cases, such moulds grew in poorly prepared compost. They were often referred to as competitor moulds or weed moulds, fungi 'growing out of place'. Some were useful diagnostic tools in that they were associated with particular compost problems, and were then called 'indicator moulds'. In this chapter we have chosen competition as the characteristic of the group, although we recognize that this may not be entirely satisfactory. The moulds included are those that, as far as is known, are not pathogens of the mushroom. Further research may show that this is not always the case, for instance the *Penicillium* spp. associated with Penicillium smoky mould. However, competition does distinguish them from the moulds (thermophilic) that play a very important part in the production of compost.

Visible mould problems are unusual in bulk-prepared compost, but yield reductions related to their presence have been reported. Fungi in 27 genera were found in a recent compost mould survey, even though the composts were generally well colonized by mushroom mycelium. Attempts to establish whether or not such moulds can be detrimental have indicated a significant reduction in yield with *Trichoderma pseudokoningii, T. atroviride,* and *Acremonium murorum.*

A small number of moulds are known to grow in well prepared phase II compost. For them to have an effect, they must become established at or just before spawning, so their presence is not so much a compost problem as an indication of lapses in hygiene which have allowed quantities of mould inoculum to be introduced at this critical time.

Trichoderma compost mould

Species of *Trichoderma* are possibly the most important moulds in mushroom culture. Trichoderma compost mould is a devastating disease which has caused considerable crop losses in Europe, North America, and Australia within the last decade. A number of species of *Trichoderma* have been shown to be pathogenic on mushroom caps, although there is less evidence that they are capable of attacking mushroom mycelium. Because of their spotting effects on mushrooms, and their likely parasitism of the mycelium, they are included in the fungal diseases chapter (*see* Chapter 4, pp. 84–89) and not here, even though there are some species for which parasitism has not been demonstrated.

Penicillium mould

Various species of *Penicillium* are associated with mushroom culture and many do not cause problems. The characteristic blue-green mould is commonly seen growing on trays, on the side-boards of shelves, on pieces of mushroom tissue left on the casing surface, and on grains of spawn. Apart from the appearance and some inconvenience, these moulds have no known measurable effect on crop yield or quality.

P. oxalicum, not associated with a specific problem, has frequently been isolated from compost. In contrast, *P. implicatum*, *P. chermesinum,* and *P. fellutanum* have been found in very severely affected crops.

Smoky mould (*Penicillium implicatum*, *P. chermesinum*, and possibly *P. fellutanum*)

Smoky mould was originally associated with *P. chermesinum*, but more recently *P. implicatum* has been found in crops with the same symptoms. Both moulds are very similar and it is possible that the original identification was not correct, or that both species are capable of causing the problem. *P. chermesinum* has been isolated from a variety of habitats, but relatively rarely, whereas *P. implicatum* is reported as being a very common soil-borne fungus. Both grow very slowly and can be difficult to isolate on agar. It is thought that the most likely source is soil on straw, or soil contamination of surfaces where compost is made. Neither species has been found on wood or on other materials within mushroom crops.

Smoky mould is characterized by a dramatic reduction in yield that, at the extreme, can be 80% or even more. Symptoms become apparent in the first flush where more or less normal mushrooms are produced at the edges of the shelves or trays, with no cropping in the centres. Characteristically, mushrooms that form around the edges open very early, producing flat caps. Often the casing on the surface of affected beds becomes colonized by other moulds, in particular Cinnamon mould (**116**). If the compost in the non-cropping areas is examined, large clouds of spores, looking like smoke, can be seen as the compost is

116 (a) Penicillium smoky mould showing edge cropping and Cinnamon mould on the casing surface. Photograph by kind permission of John Burden (b) a clear indication of a poor spawn-run in the left side trays of the crop.

disturbed. The compost has a mouldy smell. In less severe outbreaks, the first and second flushes may be reduced in yield, and by the third flush, non-cropping areas occur, together with the characteristic spore production in the compost. When the affected compost is examined microscopically, numerous penicillate sporing structures are seen. These are white in colour but turn brown with age.

The spores of both *Penicillium* spp. are slightly oval in shape and are very small with a minimum diameter of 2 microns. It is possible that some would pass through a 2-micron absolute filter.

Smoky mould has, so far, been largely associated with farms where the compost is bulk handled. Examination of compost in bulk phase II has shown the presence of the spores in the layers against the netting. When the compost is mixed, the spores are distributed throughout the compost.

It is likely that there is an association between this mould and mushroom mycelium. Maximum damage occurs when the spores are present within the compost at the time of spawning, and the mushroom mycelium is then very potently inhibited (**117**). Bulk preparation of both phase II and III is ideal for spore distribution, and as the *Penicillium* has an optimum growth temperature of 28–31°C, it is ideally adapted to compete with, and perhaps attack, the mushroom mycelium during the initial stages of mycelial colonization. The fungus will sporulate at temperatures as low as 19°C.

Controlling this mould has proved to be extremely difficult, particularly where there is a continuous

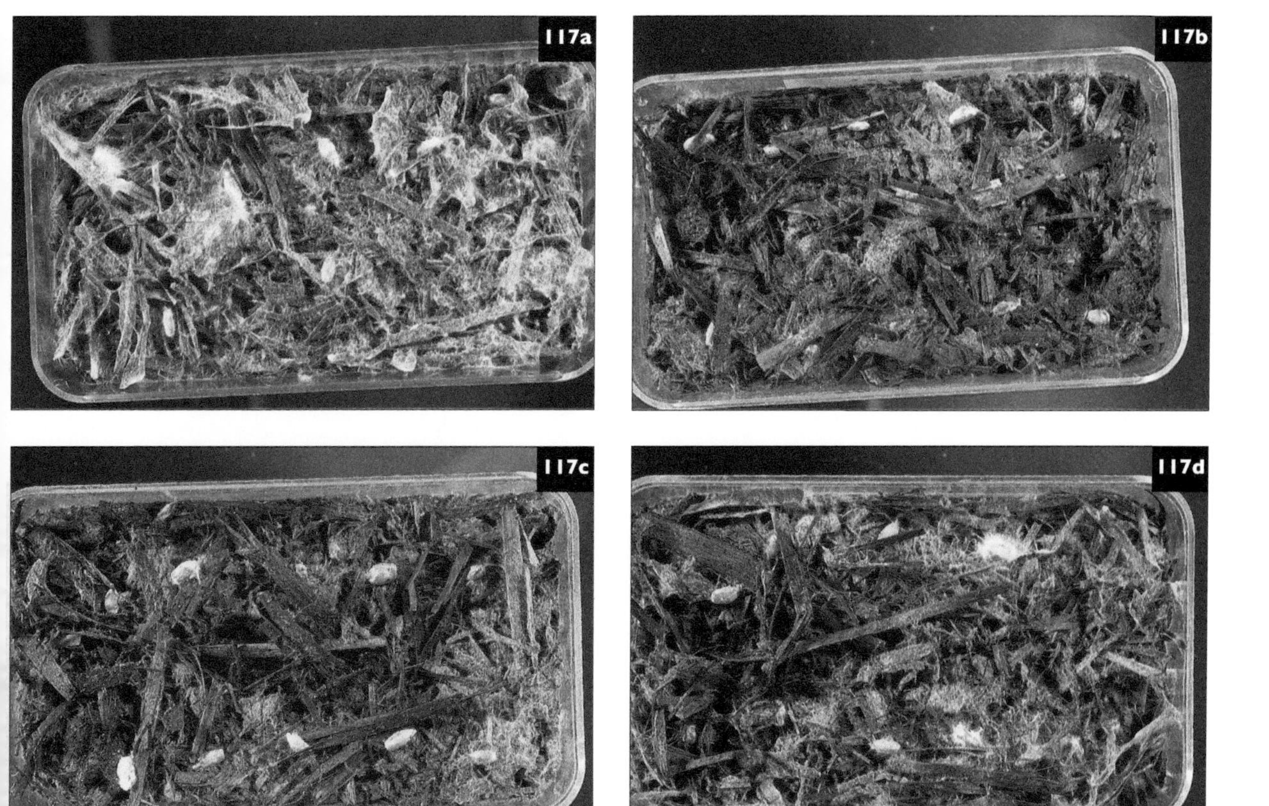

117 (**a**) Penicillium smoky mould – experimental plot spawned only; (**b**) spawned and inoculated with *Penicillium* at the same time; (**c**) inoculated with *Penicillium* two days before spawning; (**d**) inoculated with *Penicillium* two days after spawning.

flow of bulk compost around the spawning area. Fresh spores of the fungus have been killed at 47°C maintained for 10 hours or at 54°C for 60 minutes. A 2-hour treatment at 60°C was effective providing the gaseous ammonia level was high. Precise figures are not available, but experience suggests that 450 ppm at 3 hours, after maximum temperature is reached in phase II, is adequate. These thermal death-point temperatures indicate that the fungus should not survive phase II.

Filtration of the air used in phase II is obviously vitally important even though some spores may get through filters. Any small cracks or gaps within the system will add to the problem. Filters must be carefully checked to make certain that they are efficient, and replaced at the correct intervals.

The introduction of *Penicillium* spores after an effective phase II may lead to a serious problem. In modern systems, it is vitally important to separate the spawning and phase III emptying areas. The transfer of spores during emptying phase III will provide inoculum for freshly prepared phase II. Also the position of the spawning area in relation to the straw breaking area is important. If the spawning area is upwind from the straw breaking area, the chance of spore introduction at spawning is considerably increased. In order to minimize contamination at spawning, areas outside the bulk tunnels and spawning area must be regularly disinfected. Sealed spawning areas should be fumigated regularly with Formalin (*see* p. 56) to make certain that air-borne mycelial fragments and spores that have settled on ledges are killed. It is particularly important to do this every time before spawning and/or after tunnels are emptied, whether they are for phase II or bulk spawn-runs. The general hygiene principles used in the production of phase III compost should be strictly applied in order to avoid contamination of phase II compost at spawning (*see* Virus disease control, p. 116). This may include regular 'cooking-out' of phase III tunnels, especially if they are not used for the production of phase II compost.

The fungus is very sensitive to carbendazim, and control has been achieved in experiments when 50 ppm was added to the compost at spawning. There are no product recommendations for this treatment (**118**). It is also likely that the continuous use of fungicides in the compost could result in the generation of fungicide-resistant populations of *Penicillium*.

In general, the procedures described for virus disease control should be followed carefully when considering the control of Smoky mould (*see* p. 116), as the crop is likely to be most vulnerable at the same stages.

118 Penicillium smoky mould – the effect of Benlate treatment; untreated on the right and treated on the left. This fungicide treatment is not a registered use.

Pythium – black compost

At the end of the spawn-run, the compost may show discrete black areas of uncolonized compost. Black patches may vary in size from a few centimetres to a metre or more in diameter. When the compost is in trays, the patches are most frequently found in the top half, although they can extend deeper. The patches may be ovoid in shape with the upper part of the affected compost just touching the surface. In bulk run compost, the patches may occur anywhere (**119**). Eventually, the blackened areas become colonized by mushroom mycelium. Generally, crops grown from such affected compost are reduced in yield, especially in the first and second flushes. There is sometimes a higher yield in the third flush as the mycelium colonizes the previously black compost.

Examination of affected compost in the laboratory shows it to be colonized by a species of *Pythium*. In the UK and Australia, the species involved is *P. oligandrum*, a well known pathogen and antagonist of various fungi. (Polygandrom, a biocontrol product for use in the field has been developed using this fungus.) In the USA, a closely related species *P. hydnosporum* (syn. *P. artotrogus*) has been reported to be the cause.

Pythium resting spores (oospores) are very resistant to heat and drought, and have been recovered in a viable state on the surface of dry compost following phase II. In experiments, air-dried resting spores withstood temperatures as high as 90°C for 60 minutes, but in the moist state in mushroom compost they did not survive 55°C. Resting spore germination occurs at normal mushroom cultivation temperatures, although the optimum for mycelial growth is 28°C. *P. oligandrum* is able to grow quickly in well prepared phase II compost, and a growth rate of about 2–3 mm per day at 25°C has been recorded. The rate of growth is not affected by the moisture content of the compost in the range of 64–72%. A maximum inhibitory effect was shown to occur when *P. oligandrum* was introduced either before spawning, or with the

119 Pythium black compost patches in a bulk spawn-run compost. Thanks to Paul Perrin for the use of this photograph.

spawn (**120**). There was no effect on spawn growth if the *Pythium* was introduced 2 days after spawning. The duration of the inhibitory effect was 4 weeks. After this time, mushroom mycelium recolonized the affected area. This work suggests that, for a problem to occur, the compost must be contaminated at or before spawning. It seems that such a soil-borne fungus is most likely to be a contaminant of straw, either in soil splashed onto the stems, or on the roots of straw, or by storing the straw on soil before use.

Another association in this problem is with high nitrogen in the areas of compost where the *Pythium* occurs. Pockets of high nitrogen have been associated with the uneven distribution of a nitrogen supplement at spawning. It is not clear whether the areas of high nitrogen prevent mushroom mycelial growth, or whether they encourage the growth of *Pythium*. High nitrogen probably allows the *Pythium* to establish before the mushroom mycelium is able to grow. In this way it predisposes the compost towards a *Pythium* problem, although is not essential for its occurrence.

To control Black compost, it is important to avoid contamination of straw with soil either when storing the straw, or when mixing the straw on the yard. If water drains onto the mixing area from nearby fields, it is likely to contaminate the concrete with *Pythium* and possibly other unwanted organisms. In addition, care should be taken when supplementing to make certain that the supplement is as evenly distributed as possible, thereby avoiding patches with excessively high nitrogen concentrations.

In the production of phase III compost, it is important to cook-out the tunnels at intervals, and in particular to make certain that the plenum is free from dried and possibly contaminated compost.

Phenolic disinfectants have been shown to be effective in killing resting spores, and should be used as part of the clean-up programme.

120 Pythium black compost – simultaneous inoculation of the compost with *P. oligandron* and mushroom spawn. Two small agar plugs were the source of the *Pythium* (on the right-hand side of the box). Inhibition of mushroom mycelium has occurred throughout the period of spawn-run.

Other moulds

The following moulds occur from time to time and occasionally are important.

Arthrobotrys brown mould (*Arthrobotrys* spp.)
Produces brown colonies on the casing surface, not unlike those of cinnamon mould in appearance, except that they occur at any stage in cropping but tend to be more common towards the end. They are parasites of nematodes, and their presence is an indication that the nematode population of the compost and/or the casing is high. Unsuccessful attempts have been made to use this fungus for the purpose of biological control of nematodes.

Aspergillus mould (*Aspergillus* spp.)
Most frequently an indication of a composting problem. Affected compost smells musty, and generally does not support the growth of mushroom mycelium. A number of different species of *Aspergillus* have been found, including *A. niger* which appears as a black mould, *A. flavus* a mycotoxin producer, and *A. fumigatus* a thermo-tolerant fungus. The presence of the latter is a good indication of secondary fermentation and heating after spawning. *Aspergillus* spp. are very widespread in nature and capable of growing on a very wide range of materials (including leather and clothes), and some are animal and human pathogens producing a group of diseases known as aspergilloses.

Affected compost must therefore be handled extremely carefully, and wherever possible cooked-out with heat.

Black whisker mould (*Doratomyces stemonitis*)
Also known as *Cephalotrichum purpureofuscus*, it is now an uncommon mould. Its name describes the appearance of the dark-grey to black whisker-like bristles it produces on the surface of the casing. These bristles are spore-bearing structures of the fungus and may be up to 2 mm long (*see* **26 b**). *Doratomyces* is said to indicate poor initial composting which has resulted in an abnormally high compost carbon to nitrogen ratio. The fungus is thought to be antagonistic to mushroom mycelium and may affect yield. It frequently occurs with *Penicillium* and *Aspergillus* spp. Some pickers are said to be allergic to the spores. The full significance of this fungus in mushroom production is not understood.

Brown plaster mould (*Papulaspora byssina*)
Was once a very serious problem, but is now relatively uncommon. Characteristically, it appears as large dense roughly circular patches of mycelium on the surface of the casing, initially whitish but turning brown and powdery with age (**121**). It can also colonize the compost. The fungus produces numerous spores that consist of clusters of cells, often referred to as bulbils. When rubbed between finger and thumb, the bulbils feel waxy. The presence of this fungus has been associated with unconverted ammonium compounds which often result when the compost is overly wet or too broken down at an early stage in the composting process. Similar conditions favour the development of white plaster mould.

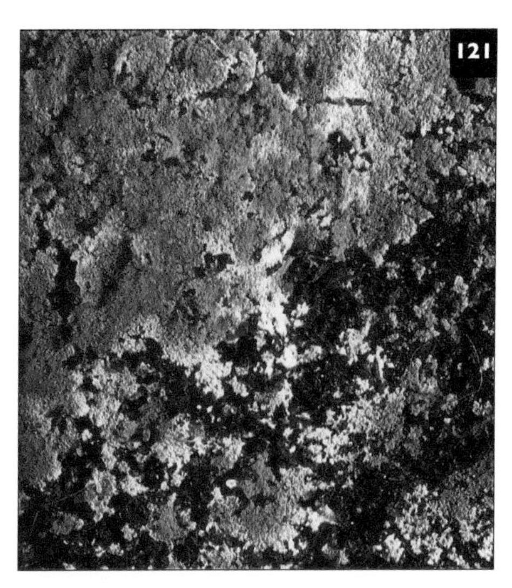

121 Brown plaster mould.

122 Cinnamon mould extensively colonizing the surface of casing. The extreme severity of this example was attributed to wet unsatisfactory compost which prevented a vigorous spawn-run.

123 Cinnamon mould – the jelly-like apothecia of the fungus often are found in the areas where the cinnamon mould was most prolific.

Cephaliophora compost mould (*Cephaliophora tropica*)

Very frequently found on dung, especially in warm countries. It is capable of growth over a very wide temperature range, from 10 to 40°C. In South Africa it has frequently been associated with poorly colonized compost.

Cinnamon mould (*Chromelosporium fulvum* teleomorph *Peziza ostracoderma* syn. *Plicaria fulva*)

One of the commonest brown moulds. It frequently occurs on the surface of the casing before the first flush. It is a very common colonizer of wet peat, which may be its main source. In tray-grown crops, patches of Cinnamon mould often occur around the legs of the trays. Occasionally, it is extensive and is then usually associated with wet and often poorly colonized compost (**122**). The colonies are frequently circular and initially are grey-white, but quickly turn brown. By the time the first flush of mushrooms has developed, the Cinnamon mould has disappeared. However, gelatinous disc-shaped circular structures (apothecia) about 1 cm across may be seen in the areas of casing where the cinnamon mould was most dense (**123**). These produce spores of the fungus (ascospores) that are shot into the air. The brown conidiospores of the cinnamon mould are also readily air-borne.

Cinnamon mould is more of a nuisance than a cause of crop loss although, where it occurs in quantity, it depletes the nutrient status of the compost, may discolour the mushrooms, and can result in a small delay in cropping, with possible slight reduction in yield.

Confetti (*Chrysosporium merdarium*)

Chrysosporium merdarium syn. *Sporotrichum merdarium* and possibly *Myceliophthora sulphurea*, is the name for a yellow mould that was common, but yellow moulds are now generally rare although serious and widespread outbreaks have more recently occurred in India (*see also* Mat disease or Vert-de-gris). Confetti mould develops in the compost and the yellow colonies of the fungus, 12–26 mm in diameter, are produced from an initial growth of white mycelium. The colonies may coalesce to form large dense yellow patches similar to mat disease. When present in large quantities, considerable yield reduction may result.

Coniophora rot (*Coniophora puteana* var. *puteana*)

A wood decay which has been reported from mushroom culture in South Africa. The fungus grows well on pine mushroom trays where it starts as white fluffy patches which spread out onto the casing layer engulfing mushrooms and pins. The affected pins turn sulphur yellow, distort and finally rot

The mould mycelium grows up the stalks of mature mushrooms turning them sulphur yellow, then yellow brown to purplish brown, and finally they collapse in a wet, soggy, and smelly mess.

Conventional tray dips are not very effective in the elimination of this fungus, although it develops less well on well preserved timber.

Fire mould (*Neurospora crassa*)
Can occur after cook-out when the crop is left *in situ* for some days following treatment. The mycelium is at first creamy-white but rapidly turns orange. Large wefts of mycelium are produced on the casing surface and often hang down the sides of beds. Large numbers of spores are produced so that, once the mould is established on a farm, it is difficult to eliminate. The problem can be totally avoided by not leaving the compost in the cook-out room for more than a day after the end of treatment.

Fusarium moulds (*Fusarium* spp. including *F. solani* and *F. oxysporum*)
Were once associated with damping-off of pins. Various species have been found in compost, sometimes in association with poor cropping. Symptoms such as damping-off, discolouration, withering, distortion, and mummification have also been linked with species of *Fusarium*, but in the absence of evidence there must be some doubt about the association of Fusarium moulds with these symptoms.

Heat mould (*Thielavia thermophila*)
A thermo-tolerant white mould which is capable of growth at temperatures well in excess of those favoured by the mushroom. For this reason, it is most frequently seen in countries where the ambient summer temperatures are high. It is a common fungus in compost, but ceases to grow when the temperature of the compost drops below 30°C. It then plays no further part in the microbiology of the compost and does not adversely affect the growth of the crop. However, if for any reason a part or the whole of the compost remains at a temperature well in excess of that required for spawn growth, this heat mould will thrive. Large white dense areas of mycelium will develop in the hot areas, and even when the temperature has been reduced the white mycelium is still visible (**124**). Other thermo-tolerant fungi may be visible in overheated compost.

The control of such fungi is entirely dependent upon accurate temperature control at all stages of compost production, but in particular at and after spawning.

Ink caps (*Coprinus* spp.)
Fruiting bodies are produced on mushroom beds, usually before cropping begins, and these subsequently disintegrate into a black slime (**125**). The mycelium of the fungus is grey and not easily distinguishable from that of the mushroom. Ink caps have sometimes occurred in large numbers before the first flush and, in spite of their presence, the mushroom crop grows well.

124 Heat mould. The white tufts of mycelium remain even when the compost temperatures are returned to their normal range.

125 Ink-caps develop quickly before cropping begins and the caps dissolve within a short time of formation. Black spores are deposited on the casing surface.

This has lead some growers to think that the presence of Ink caps can be an indication of satisfactory compost. The large numbers of spores released from the ink caps will readily, in the presence of ammonia, colonize freshly prepared compost. Their presence therefore indicates free ammonia or, ultimately, a high nitrogen status of the compost. Generally yield is not adversely affected, but Ink caps also grow well if pasteurization has been inadequate, and are then associated with very poor yields. Ink caps do not occur if compost preparation has been satisfactory.

Mat disease or Vert-de-gris (*Myceliophora lutea* syn. *Chrysosporium luteum*)

Vert-de-gris has been recognized for a very long time, but has more recently been called Mat disease because of the yellow-to-brown mycelial mats that form at the casing compost junction (**126**). Initially the colour of the mycelium is very similar to that of mushroom mycelium, and is usually only clearly distinguishable with the formation of mycelial mats.

Spores become air-borne and will cause trouble if they contaminate casing or compost. Compost ingredients, especially chicken manure and soil, are common sources. Air currents readily spread the spores. Greatest crop loss occurs when the crop is affected at an early stage. Control is usually achieved by reducing the spore inoculum levels and by strict attention to hygiene. Phase II temperatures are sufficient to kill the spores.

126 Mat disease with yellow mycelial mats at the junction of casing and compost.

Oedocephalum mould (*Oedocephalum glomerulosum*)

A fairly common mould on compost and casing. It appears as silvery grey patches on the compost during cool-down and before spawning. After spawning, the mould in the compost is light grey, but changes to fawn or light brown as the spores are produced. It sometimes covers large areas of the casing, where it is first seen as silvery grey colonies which give the appearance of droplets of dew. As the spores develop, the colonies turn brown. The spores feel gritty when rubbed between finger and thumb. The effect of this mould on yield or quality is not known. It is said to indicate the presence of ammonia and amines and in this respect it is probably an indicator of unsuitable compost.

Olive-green mould (*Chaetomium globosum* syn. *C. olivaceum*)

Was a common mould, but with improved methods of composting is now seen far less frequently. The mycelium is grey-white and is often mistaken for that of the mushroom during the early stages of spawn-running. When extensive in compost, it reduces yields in proportion to the extent of the colonization of the compost. It does not grow in casing. Compost with Olive-green mould is often black, and is not colonized by mushroom mycelium. Some mushroom growth may eventually occur in affected areas, but only sparsely and not enough to give a satisfactory crop. If Olive-green mould develops shortly after the end of phase II, there is little hope of achieving a satisfactory yield.

The fungus is usually identified by the olive-green fruiting structures, which are about the size of a pinhead and have a rough spiny appearance. These are produced in large numbers on the straw, often well spaced, and clearly visible. As they age, they darken in colour and eventually become brown (**127**). *Chaetomium* spores are able to withstand higher temperatures and higher concentrations of ammonia than mushroom mycelium. They are present in compost ingredients and their development is favoured by anaerobic conditions during phase II. It is reported that oxygen levels of 16% or less are conducive to the development of *Chaetomium*. Such conditions may occur when the compost is too wet, over-composted, overheated (compost temperature greater than 62°C in phase II), not adequately aerated during phase II, or over-compacted at filling. Good composting in phase I and control of the environment during phase II will prevent Olive-green mould, but there is little that can be done to save affected compost.

Pearly white plaster mould (*Botryotrichum piluliferum* teleomorph *Chaetomium piluliferum*)

Very similar to White plaster mould in appearance, except that the colonies have a pearly glisten, and the mycelium is creamy white to buff, instead of white. The effects of this mould are similar to those of other plaster moulds.

Pink flour mould (*Trichothecium roseum*)

Has occasionally been reported, particularly in the compost adjacent to wooden boards. It is initially white but develops a pink colour as spores are produced. There are reports of it infecting mushrooms in India, and in the Czech Republic it was associated with brown sunken lesions on mushroom caps.

Red geotrichum or Lipstick mould (*Sporendonema purpurescens* previously known as *Geotrichum purpurescens*)

Is found in compost ingredients. The spores germinate to produce a fine white mycelial growth, not unlike mushroom mycelium, particularly on the surface of the casing, but also sometimes in the compost. Its colour generally changes to bright pink and finally buff, with a powdery appearance as the spores are produced (128). The large numbers of spores are easily air-borne. Where outbreaks are severe, the farm becomes extensively contaminated by the spores, which are a source of inoculum for succeeding crops.

The presence of this mould in the crop has a marked inhibitory effect on the growth of mushroom mycelium and is often associated with crop loss, particularly when present in the compost before casing. Contamination of the compost at cool-down or during spawn-run is generally most damaging.

It is important to achieve an evenly high temperature during phase II (*see* p. 10) in order to minimize the risk of lipstick mould developing. Generally, an effective phase II and strict attention to hygiene are the best approaches to control.

Roumegueriella mould (*Roumegueriella rufula* syn. *Lilliputia rufula*)

Has been reported from the UK, but is better known in India. The white mycelium and round fruiting bodies with a pointed apex are seen in affected compost. It is believed that animal manures are the source. The mould is able to compete with mushroom mycelium, inhibiting spawn growth, which results in a reduction in yield. Poor pasteurization is the probable initial cause.

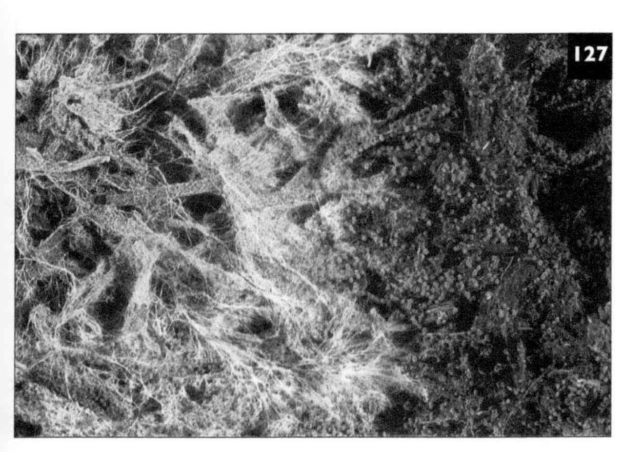

127 Olive-green fruiting bodies of *Chaetomium* inhibiting the growth of mushroom mycelium.

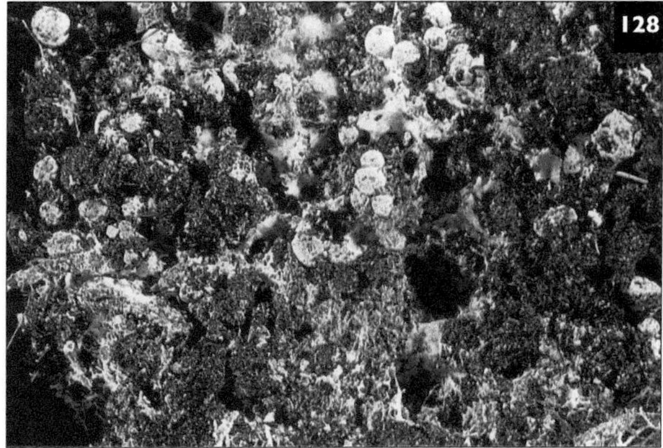

128 Lipstick mould seen on the casing surface. The pink mycelium of the mould is visible between lumps of casing.

Sepedonium yellow mould (*Sepedonium* spp.)
Uncommon. It develops in the compost and although initially white turns yellow to tan with age. Numerous spores are produced that are readily air-borne, and these can contaminate compost during preparation. The large spherical spores are said to be very heat-resistant. It has recently been found in the basal layers of compost produced by bulk methods, and also in the bottom of troughs. Its presence in trough culture is thought to have caused distorted mushrooms, possibly due to the production of a volatile toxin.

Sepedonium affects mushroom growth in well prepared compost. The most severe effects result from contamination before casing. There are reports from South Africa of *S. niveum* causing distortion of mushrooms not unlike those caused by *Mycogone perniciosa*.

Vern Astley disease (*Oidiodendron sindenia*)
Was originally associated with a species of *Spicaria*. Vern Astley disease has been reported on many occasions but is now rare. The symptoms closely resemble those described for Penicillium smoky mould (*see* p. 128). Although called a disease, there is no evidence as yet that the fungus associated with the symptoms is parasitic on mushroom mycelium, but it does compete very strongly for nutrients and space. The mycelium of *Oidiodendron* closely resembles that of the mushroom, not only because of its colour, but also its growth habit in the compost. When present in large quantities, it sporulates profusely, resulting in a smoky mould symptom when the compost is disturbed. It has been linked with respiratory problems of workers. Affected compost may have a slight chlorine odour. Other symptoms of this disease include early opening of mushrooms, spongy tissue at the base of the stalk and flattened scaly caps.

Experiments have shown that *O. sindenia* colonizes compost more readily when mushrooms are being produced, rather than during spawn-run. The fungus is very slow growing and therefore takes time to build up to damaging concentrations. Spores can enter the compost during spawn-run, or with the casing. They have an optimum germination temperature of 24°C. The earlier the introduction, the greater is the effect on the crop. Because of the very slow rate of growth, symptoms may not become apparent until the second flush. At an advanced stage of this disease, the mushroom mycelium appears to die out in the worst affected areas.

Control is achieved by the prevention of the entry of spores, particularly during the early stages of crop production. The spores are only slightly more than 2 microns in diameter, so it is very important to have the correct filtration equipment to prevent their entry. Spore contamination is likely to be widespread on an affected farm, and thorough hygiene is essential in the elimination of the problem.

129 White plaster mould.

White plaster mould (*Scopulariopsis fimicola*)
Still seen fairly frequently. As the name suggests, dense white patches of mycelium and spores occur on the casing surface and in the compost (**129**). Many other weed moulds initially produce white mycelium, but they change colour as they age, whereas *S. fimicola* remains white. The fungal growth is sometimes so dense that it looks as though flour has been put on the surface of the bed. White plaster mould may reduce yield by competing with mushroom mycelium, or it may have little or no effect on yield. It grows particularly well in compost with a high pH (8.2 or above is stated in the literature).

Action points for control of mushroom moulds

- Identify the problem. The correct identification of the mould may give a clear pointer to the most effective method of control. Composting problems are the most frequent causes of moulds.
- Generally, when moulds develop in well prepared compost, it is important to identify the sources and then take the necessary measures to eliminate them.
- The development of moulds that are known to be associated with particular compost problems, e.g. Olive green mould, are a clear indication that compost preparation needs investigating.
- Strict hygiene throughout the farm is necessary to reduce the general level of spores and mycelial contamination (*see* Chapter 3).
- Temperatures during all stages of compost and crop production must be checked and equipment recalibrated if required. Achieving the correct temperature during all stages, particularly in phase II, is very important.
- Ammonia levels in phase II should be at least 450 ppm 3 hours after the maximum temperature has been reached.
- All equipment must be thoroughly cleaned between operations.
- Periodic tests should be done on apparently healthy phase III compost in order to monitor for the presence and build-up of unwanted moulds.

CHAPTER 8

Pests

This chapter was originally written by P.F. White, formerly of Horticultural Research International, and has been revised by the current authors with help from Jane Smith of Warwick HRI.

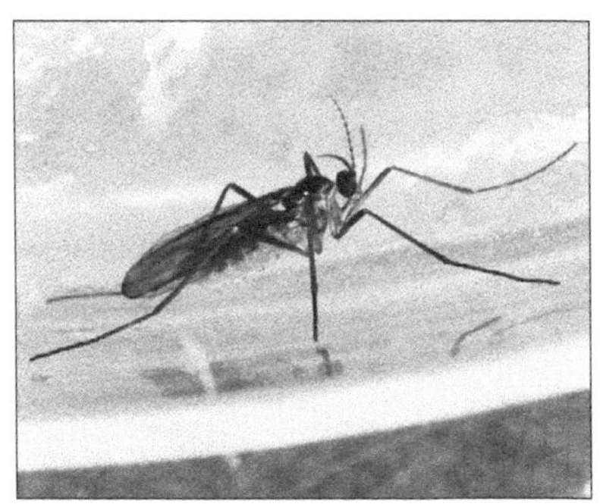

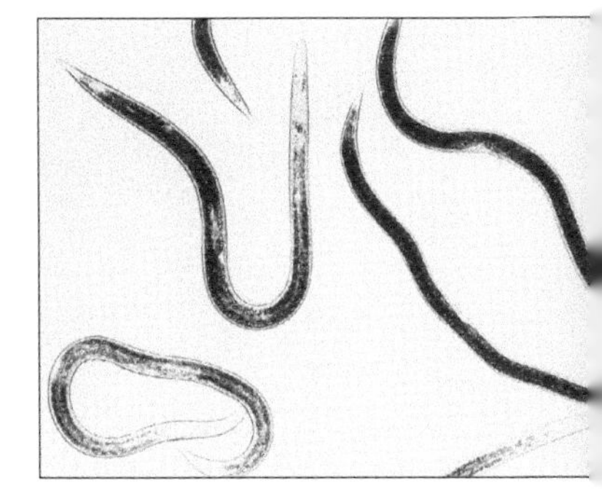

Introduction

Mushroom production is affected by a number of pests capitalizing on the favourable environment and the bountiful and continuous supply of food available to them.
The fauna associated with phase I composting is extensive, but will not survive an efficient phase II pasteurization. It follows that, for subsequent crop loss to occur, pests must colonize a crop after the pasteurization process. As the initial infestation level of any flying pest is likely to be small enough to escape casual observation, it is good management practice to use a monitoring technique to determine the relative abundance of such pests (*see* p. 38).
When a potential pest is first noticed, it is extremely important to identify it correctly, as control regimes for the various pests differ in enough respects to cause control failures. It is also important to understand how pesticides act. For instance, some are poisons and likely to act quickly, while others affect physiological processes and are slower to act. The few biocontrol products that are available depend upon parasitism.
The choice of pesticides for the control of mushroom pests varies considerably from country to country and from time to time. It is therefore vitally important to check the national list of registered products before selecting one.

Mushroom flies

The most serious pests of cultivated mushrooms belong to a group of insects called the Diptera, so-called because the adults are two-winged flies. As the various pest species are similar in size and colour, they are often referred to as 'flies', a general term which belies the importance of accurate identification for sound control practices. *Table 6* shows a comparison of the distinguishing features of the three main groups.

TABLE 6 A summary of the distinguishing features of the three main fly pests of the cultivated mushroom

	Sciarid	Phorid	Cecid
Adult fly	Largest fly pest (3–6 mm) Recognizable by prominent antennae Lurks on casing and walls. Moves quickly but smoothly.	Characteristic hump-backed appearance (2–3 mm). No visible antennae. Fast jerky movement, attracted to light.	Very small (1.5 mm) and rarely seen.
Larvae	Distinguishable by distinct black head. Size up to 12 mm. Occasionally seen in poorly colonized compost but most commonly in hollowed-out pins or small buttons. In heavy infestations may be found in tunnelled mushrooms.	Larvae smaller than sciarid (1–6 mm) Only rarely seen, except in very heavy infestations resulting in mycelial and pin destruction.	Characteristic orange or creamy white larvae visible in swarms on mushrooms.
Vulnerable infestation stage	Phase II cool-down, early stages of spawn-running and before casing colonization.	Growing mycelium, i.e. spawn and case-running.	A contaminant pest introduced in poorly pasteurized compost, tray timber, casing, very occasionally by adult flies. Larvae sticky and easily disseminated around the farm.

Sciarids or fungus gnats

A number of species of sciarid have been recorded as infesting mushroom crops in Britain. Intrinsic difficulties in their identification have led to some confusion in the past. Two species predominate in many countries where mushrooms are grown, *Lycoriella castanescens* (syn. *L. auripila*) and *L. ingenua* (syn. *L. solani*, *L. mali*). *L. castanescens* causes more primary damage to mushrooms than *L. ingenua*. Both species are commonly found in the UK, while in the USA *L. ingenua* is most frequently reported. *Bradysia* spp. have also been associated with mushroom crops in some countries.

Information gathered from trapping studies indicates that sciarids tend to be on farms all year round, in contrast to phorids which are more seasonal pests, a factor that may have contributed to the development of insecticide resistance in sciarids.

Symptoms

Sciarids are able to affect all stages of post phase II mushroom production. A heavy infestation by flies at or before spawning may inhibit the spawn-run because the larvae in the compost produce large quantities of faecal matter, resulting in localized areas of foul soggy compost. The mushroom mycelium will be unable to colonize this contaminated material, and poor yields will subsequently result.

The most frequent damage caused by sciarid larvae is tunnelling in the stalks, seen when the mushrooms are harvested (**130**). Stalk tunnelling which can be extensive – appears to occur predominantly – but not exclusively, when the population of sciarids is large.

Larvae feeding on developing pin-heads and buttons can cause serious damage (**131**). The mycelial attachments can be severed, causing the pin-heads to

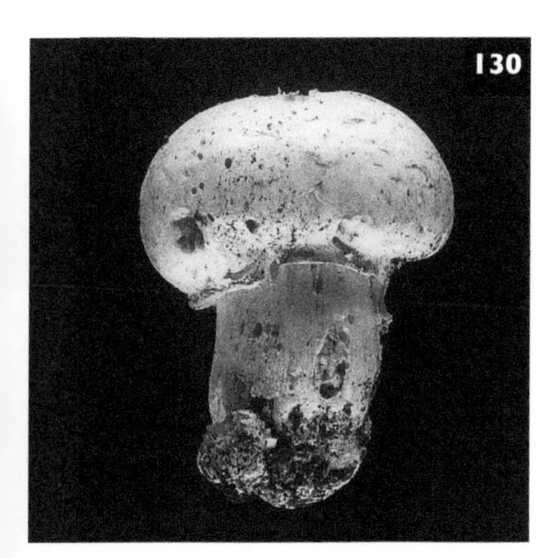

130 Sciarid larval feeding holes in a mature mushroom.

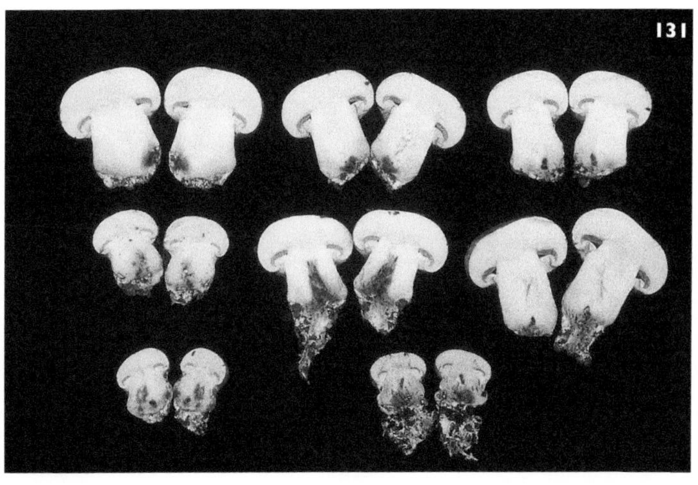

131 Sciarid larval damage to mushrooms. Small mushrooms are almost completely colonized and there is also extensive damage to marketable mushrooms.

become brown and leathery; or the pin-head/button can be hollowed out producing a sponge-like mass, or may be consumed entirely. This sort of damage occurs at a lower population density than is required for stalk tunnelling, and can go unnoticed by the grower.

The adult flies are also capable of causing damage to a crop, not directly, but as a result of contaminating pre-packed mushrooms.

Various mite species are able to breed in the foul areas of compost and decaying pin-heads created by sciarid larvae. The mites are able to cling on to adult sciarid flies, and as many as 30 per fly have been recorded. As these mites are often associated with different bacterial diseases, and *Verticillium* spp., their dispersal around a mushroom farm may result in pathogen transmission. In a similar way, the adult flies are capable of spreading the spores of *Verticillium* spp. and probably other pathogens such as *Cladobotryum* spp. have also been said to be spread by flies, but their spores are not sticky and it seems unlikely that this means of spread is significant. Adult flies are frequently a very important means of pathogen transfer between crops.

Description of pest

Mushroom sciarids are small (3–6 mm long), delicate, black gnat-like flies with large compound eyes and long thread-like antennae, which are held characteristically erect (**132**). The abdomen of the female fly is larger than that of the male, as it is

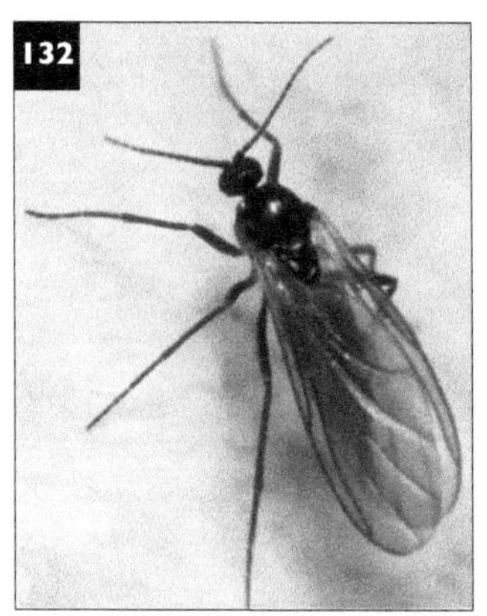

132 Mushroom sciarid (*Lycoriella castanescens*) adult female.

133 Sciarid fly – abdomen of a male showing prominent claspers at its tip.

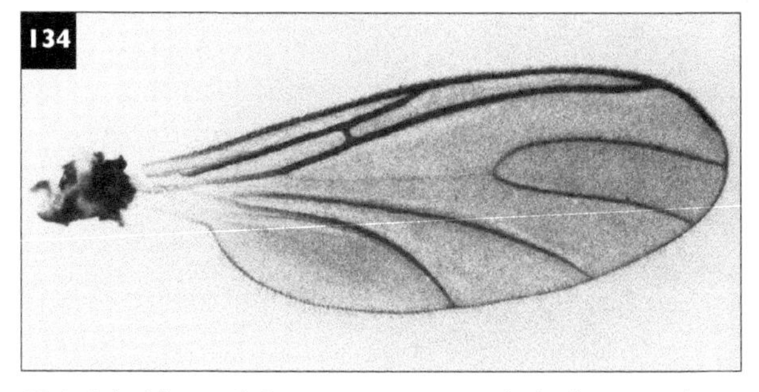

134 Sciarid *Lycoriella castanescens,* typical wing venation.

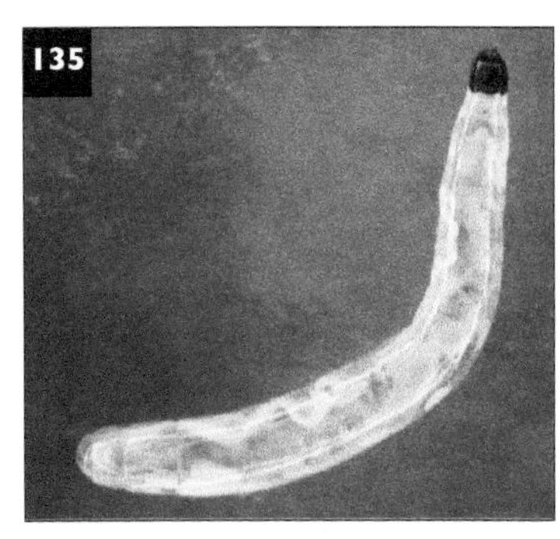

135 Mushroom sciarid larva (*Lycoriella castanescens*) with the characteristic black head.

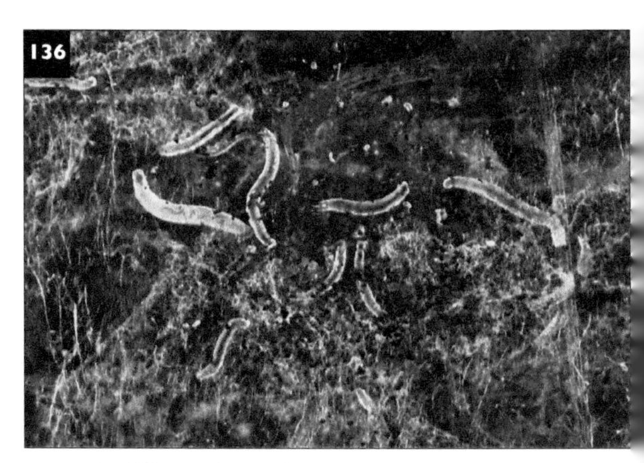

136 Sciarid larvae in mushroom compost.

generally bloated with eggs, and its apex is quite pointed. The male abdomen is quite slender and terminates in prominent genitalia with well-developed claspers (**133**). The characteristic Y venation in the iridescent wings is also an important identification feature of sciarids (**134**).

The larvae are white, legless, fairly active maggots which, when mature vary from 6–12 mm in length. The main identification feature is the distinct large head, which is black and shiny (**135, 136**) and bears large powerful chewing mouthparts. Very occasionally larvae can be found suspended between mushroom beds on fine silken threads. This may be a means of migration to other areas within the crop.

Epidemiology

The adult flies are not as active as phorids near lights. Females tend to rest on the surface of trays and walls, while males often remain on the surface of the casing, in readiness to mate with the newly emerging females. In their natural habitat, sciarids inhabit leaf mould, wild fungi, rotting wood and vegetable matter, and it is from these sources that they originally infest mushroom farms.

Sciarids that infest during phase I composting will be killed by the high temperatures reached during the phase II process. Populations that emerge during cropping must be the result of infestations that occur after this process. These infestations arise from sciarids that are attracted to the fermentation odours being given off during the cool-down period of phase II. Although the farm is the most likely source of flies, bought-in phase II compost and, in warm weather, nearby damp organic matter, can also be sources. Adult flies are sometimes introduced onto a farm with mushrooms imported from other farms. Whether or not this means of introduction is important is dependent upon the farms layout and marketing practices.

Females can lay up to 170 eggs. It has been shown that they tend to favour compost that has not been colonized by mushroom mycelium, and even a small amount of mycelial growth reduces the numbers of eggs laid. The rate of development, from egg to adult, varies considerably with the compost temperature. At a temperature of 24°C, it takes only 18–22 days. During cropping, compost temperatures are generally cooler and generation times are about 35–38 days at 18°C. At 10°C the total time is nearer to 50 days. This range of temperatures explains why sciarids can emerge at any period during cropping, although the majority would probably emerge 21–25 days after spawning.

Loss of yield is proportional to the average number of larvae present throughout the cropping period. A mean of just one larva in a handful of casing causes 0.5% loss in total yield. This figure is said to represent the economic threshold for this pest based on the cost of the control measures on an average farm. Another cause of loss in modern systems of marketing is the presence of adult flies in wrapped pre-packs, which can result in mushrooms being rejected by the retailer.

Control

Access of sciarid adults to phase II, and spawn-running rooms can be prevented by the use of filters and screening on ventilation ducts (mesh size not greater than 0.3 mm), and efficient sealing around doors will also help to keep them out. Modern systems of compost production where air is either recirculated or filtered have greatly reduced sciarid problems. The number of flies that get past these physical barriers should be monitored with the use of sticky traps, which are examined daily.

Systems in which phase II tunnels are emptied into unprotected areas, or the compost filled over an extended time period onto shelves, can be very vulnerable to fly infestation.

An antagonism exists between sciarids and mycelium such that large volumes of mycelium inhibit larval development, and well-colonized compost does not favour egg-laying by females. Thus a vigorous spawn-run throughout the compost will, by deterring sciarid egg-laying, exert a degree of cultural control. Conversely, a poor compost through which mycelium has difficulty in growing will provide an attractive substrate for sciarid development.

Insecticides that can be used for fly control vary from country to country so it is important to consult the national list of registered products before making a choice. In many countries, there are no registered insecticides that can be added to the compost. However, in the USA, and Canada cyromazine (Armor or Citation), can be mixed into phase I compost, preferably one turn before filling into phase II; and azadirachtin (Amazin, Azatin) and methoprene (Apex, Sciariprene), growth regulator insecticides, can be added to phase II compost before spawning.

In Australia, triflumuron (Alsystin) is registered as a larvicide for use in the compost or in the casing, while fipronil (Regent) can be mixed into the casing.

The organophosphorous (OP) materials chlorfenvinphos (Birlane) and diazinon can be used in the compost for both sciarid and phorid control in the few countries where they are registered. OP resistance is common in some sciarid populations, and then OP insecticides are no longer effective. There are some records, for instance in the USA and the UK, of sciarids developing resistance to insecticides that are not OPs, e.g. diflubenzuron (Dimilin).

Newly spawned crops should be protected from flies entering in the first few days. This can be done by applying bendiocarb (Ficam) to walls, doors, around doorways etc together with aerosols or fogs of pyrethrins. To control larvae, casing can be treated with the parasitic nematode *Steinernema feltiae* in such products as Entonem, Nemacel, Nemasys or Scia-rid or the insecticide diflubenzuron. Nematodes target larger larvae and are typically applied after casing. To gain maximum effect from diflubenzuron or cyromazine, it should be incorporated during the preparation of the casing. Diflubenzuron or cyromazine can also be used as a post-casing drench, but to ensure maximum penetration, drenching must be done as soon as possible after casing. Casing can also be drenched with *Bacillus thuringiensis* var. *israelensis* (Gnatrol, Vectobac). This bacterium has a toxin which is poisonous to sciarid larvae. Application to the casing should be timed so as to maximize exposure to young sciarid larvae. Methoprene (Apex) can be incorporated or drenched onto the casing. In this case, the application needs to be timed for the presence of the third instar sciarid larvae. Both *B. thuringiensis* var. *israelensis* and methoprene have a relatively short life in the casing compared with diflubenzuron and cyromazine.

Flies emerging before or after cropping, usually from just after casing onwards, should be controlled by aerosols or fogs of pyrethrins. In the UK, there is an off-label recommendation (at growers' risk) for the use of deltamethrin (Decis) for this purpose.

Action points

- Use effective filtration of ventilation ducts in the phase II and spawn-running rooms.
- Expose phase II compost at tunnel emptying and shelf filling for as short a time as possible.
- Use knock-down sprays during the first week of spawn-run and the first week after casing.
- Use dust filters on the exhaust vents to minimize the distribution of flies around the farm.
- Use sticky traps to monitor flies in the spawn-running room and generally throughout the farm.
- Use insecticides, where available, in phase I compost.
- Incorporate diflubenzuron or cyromazine in the casing or drench immediately after casing; alternatively use a nematode product.
- Use knock-down sprays or fogs during cropping.
- Treat surfaces in the cropping house and around the farm with bendiocarb.
- At the end of cropping ensure that an effective cook-out temperature is reached (*see* p. 46).
- Make sure that spent compost is removed from the farm.
- Minimize breeding areas around the farm by removing excess organic debris and by draining nearby wet land.
- Do not store mushrooms from other farms close to cropping houses.
- For a precautionary note on the use of chemicals *see* p. 6.
- It is vital to check the National list and read the product label carefully before using any of the pesticides mentioned

Phorids

The most important species is *M. halterata*, both in Europe and North America. Other species affect wild mushrooms (*see M. nigra*).

Megaselia halterata

Symptoms

M. halterata larvae feed solely on mushroom mycelium and in this way are able to cause a reduction in yield. As the larvae do not burrow into mushrooms, and develop mostly in the compost, they are seldom seen. It is the flies that are the most obvious and these are normally numerous in summer and late autumn. They are very active near lights and can be a considerable nuisance to pickers. They are also important vectors of *Verticillium* spp. (*see* p. 73). Individual flies are capable of spreading large numbers of spores. With present-day fly control, phorid flies, by acting as fungal vectors, pose a much greater threat to the mushroom crop in this way than they do as a result of their own feeding damage.

Mushroom crops affected by flies, particularly phorids, often have a characteristic smell which is reminiscent of dilute urine.

Description of pest

Mushroom phorids are small (2–3 mm), hump-backed flies with inconspicuous antennae (**137**). They resemble diminutive house flies, are brown-black in colour, and generally stouter in appearance than sciarids. There is no obvious difference between the male and female flies although, on closer examination, the male abdomen ends in a black capsule, while the female abdomen is generally paler and its apex is pointed. The wing venation is also a characteristic identification feature (**138**).

The larvae are creamy-white legless maggots (1–6 mm long) with a pointed head end which is not black, and a blunt rear end (**139** *cf.* sciarid larva, **135, 136**). Two-thirds of their immature life is spent

137 Mushroom phorid (*Megaselia halterata*) adult fly.

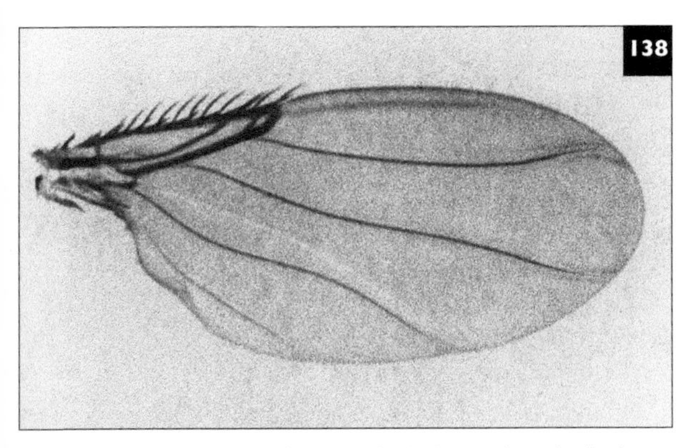

138 Mushroom phorid (*Megaselia halterata*) typical wing venation.

139 Mushroom phorid larvae.

as an immobile non-feeding pupa (**140**) 2–3 mm long, varying in colour from creamy-white to brown as the fly inside develops. They can sometimes be seen in the compost, especially at the sides and corners of the boxes. Each female is capable of laying about 50 eggs which she places within a millimetre of growing mycelium. Female phorids may live several weeks after egg laying, and in cool weather this may extend to as much as 10 weeks.

Epidemiology

M. halterata is unable to fly when the air temperature is below 13°C, so populations from the wild are unlikely to invade a mushroom farm in the winter. Short warm spells can result in isolated infestations. This highlights the importance of continually monitoring numbers of flies invading a farm. Above 17°C, temperature is no longer a limiting factor.

Windy weather curtails flight, but not crawling, while rain needs to be quite heavy to prevent activity because in still conditions, flight may occur in drizzle or even light rain. Phorids do not fly after twilight, no matter how high the temperature.

Male flies congregate on the doors of cropping houses and spawn-runs, and the females make directional flight towards them. Once mated, the females become attracted to the odour emanating from growing mushroom mycelium. They are able to detect small volumes of spawned compost even at a considerable distance. Phorids fly upwind to the source of the aroma, and lay their eggs. The compost is attractive from the day of spawning until the casing is fully colonized, reaching a maximum attractiveness at casing.

Within the cropping house, the flies congregate on the surface of boxes or shelves, often at the top corners of a stack of trays, and especially near lights or doors. When seen on compost or other surfaces, they scuttle about with a characteristic rapid jerky run.

The rate of development from egg to adult varies considerably with the compost temperature. At a temperature of 24°C it takes only 17 days for the adult to develop. During cropping, compost temperatures are generally lower and at 18°C generation times are about 45 days. This range of temperatures explains the reason phorids can emerge at any period during cropping, although the majority would probably emerge 3–4 weeks after spawning.

The economic threshold for larval density has not been determined, although it appears that many more phorid than sciarid larvae are required to cause significant crop losses.

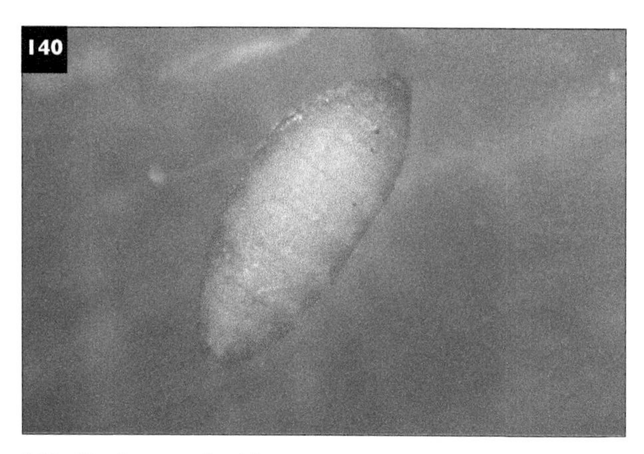

140 Mushroom phorid pupa.

Control

The use of bulk compost, in particular phase III, has made invasion more difficult, and largely has eliminated phorids from such farms. This clearly illustrates the importance of the exclusion of phorids during spawn-run. On conventional farms, it is very difficult to achieve the same degree of exclusion. Pre-cropping periods from casing onwards must also be protected whenever conditions are suitable for flight. Aerosols and fogs of pyrethrins can be used, although their effects are very short-lived. The short duration of protection given by these materials is a great disadvantage if flies are continually entering a building during suitable 'flight' weather. Bendiocarb (Ficam) can be used on surfaces in some countries and can be effective for longer periods than the pyrethrin aerosols. If fresh air is used for cooling purposes, it is unlikely that any sort of aerial spraying system will be effective.

In any attempt to exclude flies, by physical means, from the spawn-running and case-running rooms, careful attention should be given to screening ventilation ducts, or any point where air is coming out of the house, as this will carry smells that attract the flies. Because the females are able to squeeze through very small cracks and crevices, the screening should have a fine mesh (maximum diameter 0.3 mm). The sealing around screens, doors, and cracks in walls should be exceptionally thorough.

Water drip screens, in which droplets of water are constantly dripped from an overhead pipe at the house entrance, have been used successfully on some farms.

Insect growth regulators such as diflubenzuron, cyromazine, and methoprene, and biological control agents such as *Bacillus thuringiensis* var. *israelensis*, and nematodes, are relatively ineffective against phorids.

Diazinon, mixed into the compost, controls phorids in those countries where its use is permitted.

Flies emerging during cropping can be controlled by fogs or aerosols of pyrethrin-related compounds, although routine use of such compounds can result in resistance. It cannot be over-emphasized that the control of flies during cropping is merely a cosmetic operation in terms of fly control, but it can be very important in reducing the spread of pathogens as well as reducing numbers of flies invading new crops

Action points

- Thoroughly seal gaps or cracks in doors and walls of spawn-running room.
- Use effective screening of ventilation ducts in spawn-running rooms and case-running rooms.
- Use pyrethrins to kill flies during cropping.
- Paint doorways, walls and other surfaces with bendiocarb.
- Use sticky traps to monitor the number of flies in the spawn-running room.
- At the end of cropping, ensure that an effective cook-out temperature is reached throughout both the growing medium and the trays or shelves.
- Make sure that all the spent compost is removed from the farm.

Megaselia nigra

Although the black adults of this phorid are frequently seen on wild mushrooms, it is an uncommon pest of the cultivated crop.

Symptoms

Megaselia nigra larvae feed on, and in, developing mushrooms. The cap and the stalk can be tunnelled by numerous larvae (**141**) and, in extreme cases, the mushroom may collapse as a result of extensive tunnelling which is often followed by bacterial decay. In contrast to sciarid tunnelling, *M. nigra* always starts from the top, while sciarids tunnel from the base and rarely colonize the cap (*see* **130, 131**). Burrowing by *Drosophila* larvae can also affect the cap (*see* p. 155), but this pest is uncommon.

Description of pest

The adults of *M. nigra* are slightly larger and darker than those of *M. halterata*, but otherwise have the same general characteristic appearance. In nature, the adult female lives for 16 days and the male for 10. The larvae of both species are similar, except that those of *M. nigra* are generally longer and possess a pair of distinct black 'mouth-hooks' on their pointed heads, with which they are capable of burrowing within mushroom tissue.

Adult females usually lay about 50 eggs attached to the gills of developing mushrooms, although they can also be laid on the casing surface or immediately next to the developing pin-heads. At 18°C the eggs hatch in 3 days, and in only a further 5 days become pupae. The fly emerges from the pupa after 5 more days.

Epidemiology

M. nigra can readily be found in wild mushrooms, where tunnelling is very common, especially in the warmer late summer weather. It is now an uncommon pest of commercial mushroom growing. Light traps have shown that adults fly from mid summer to mid winter, and that they are likely to be troublesome in late summer. *M. nigra* lays its eggs only where there is natural daylight. In most modern mushroom houses, it is very unlikely to be a problem except occasionally when doors are left open for cultural operations or for harvesting.

Control

To prevent egg-laying, ensure that the mushroom beds are not exposed to natural daylight.

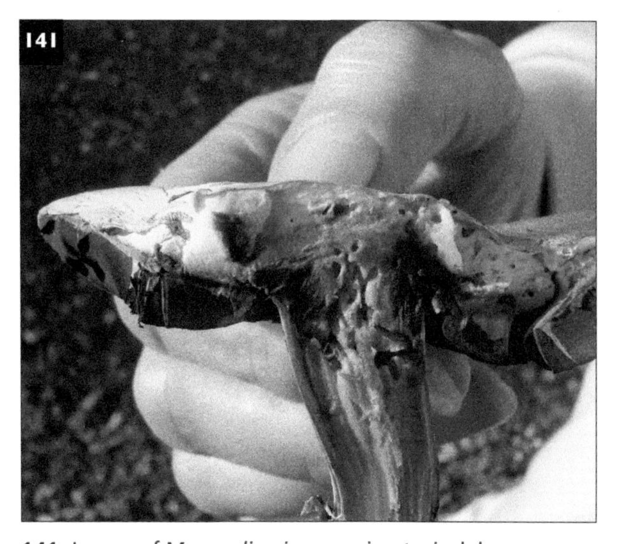

141 Larvae of *Megaselia nigra* causing typical damage.

Action point

- Black-out any openings in buildings which allow natural daylight to fall on the mushroom beds, and be aware of the risks of leaving the house doors open for cultural and harvesting operations.

Cecids

A number of different cecids have been recorded on mushrooms. Some are paedogenetic (mother larvae give rise to daughter larvae), others are not.

Paedogenetic cecids

Heteropeza pygmaea and *Mycophila speyeri* are the commonest and are economically important. A third species *M. barnesi* has also been found but, as it develops more slowly than the other two, it is generally not seen until the third or subsequent flushes.

Symptoms

The white (*H. pygmaea*) or orange (*M. speyeri*) larvae are normally first noticed when, after watering, they swarm on to mushrooms and sometimes the bed surface and floor (*H. pygmaea*). Larvae feed on the outside of stalks or at the junction of stalks and gills (**142**). Loss in marketable yield is largely due to spoilage. With *H. pygmaea*, but not *M. speyeri*, the larvae carry bacteria which cause brown discoloured stripes on mushroom stalks and gills. The delicate gill tissue subsequently breaks down to produce tiny drops of brown-black fluid (**143**).

The presence of larvae in marketed packs of mushrooms can result in the rejection of the whole of the product with considerable financial loss.

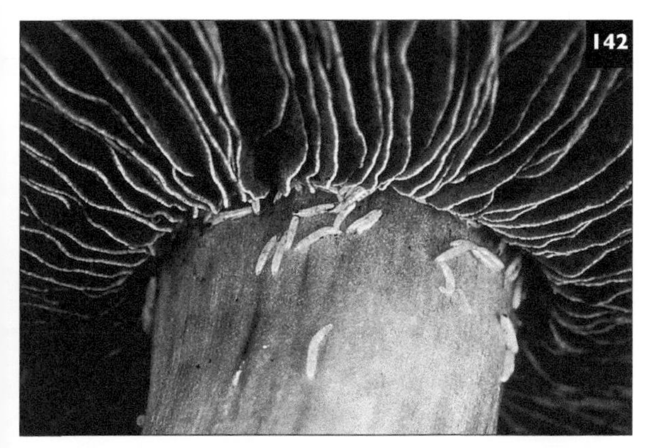

142 Larvae of the mushroom cecid, *Heteropeza pygmaea*, feeding on a mushroom.

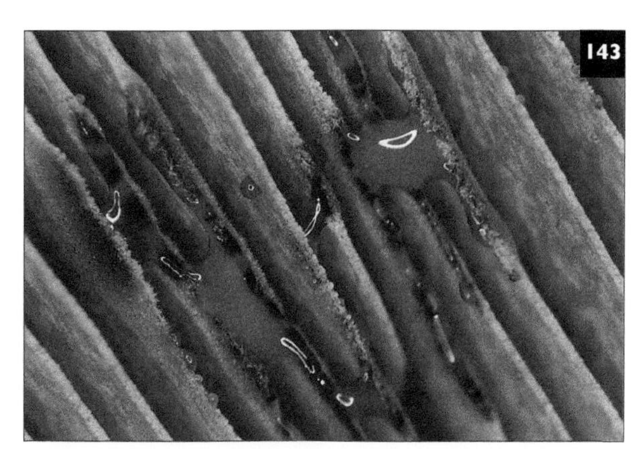

143 Mushroom cecid damage to mushroom gills rapidly colonized by bacteria causing a gill rot. The symptoms are very similar to those of Drippy gill.

Description of pests

Cecids are rarely identified from the fly stage, because they are minute (about 1 mm) and seldom seen (**144**). The larvae are the main means of identification. The maggots of *H. pygmaea* are white and 1.0–2.8 mm in length whereas the maggots of *Mycophila speyeri* are orange and about 0.8 mm long; these lengths are however variable (**145, 146**). They have no discernible head, but there are two 'eye-spots' at the head end, which give the appearance of an 'X'. Although legless, the larvae move by flexing and straightening their bodies, which enables them to jump several centimetres. In dry conditions they become tacky and then stick together in large masses often 3 cm across. The larvae tend to move towards sources of light so are most frequently seen along main gangways, near doors and other light sources. Within compost, larvae are able to puncture the mycelium and suck out cell contents. Large larvae may tear tufts of mycelium and then feed on the exuding sap.

Reproduction is unusual in that it is normally achieved by paedogenesis. Each cecid larva becomes a 'mother larva' which will give birth to about seven (*H. pygmaea*) or 20 (*M. speyeri*, **147**) 'daughter' larvae within 6 days or so of its own birth. Because several stages of a normal reproductive life cycle are bypassed, this method of reproduction leads to very rapid multiplication. Consequently, enormous populations, as large as 18,000 per handful of casing, can develop.

144 Mushroom cecid, *Heteropeza pygmaea;* female fly. Reproduced by kind permission of Dr I.J. Wyatt.

145 Mushroom cecid, *Mycophila speyeri;* three larvae showing their range in size and distinctive orange colour. Reproduced by kind permission of Dr I.J. Wyatt.

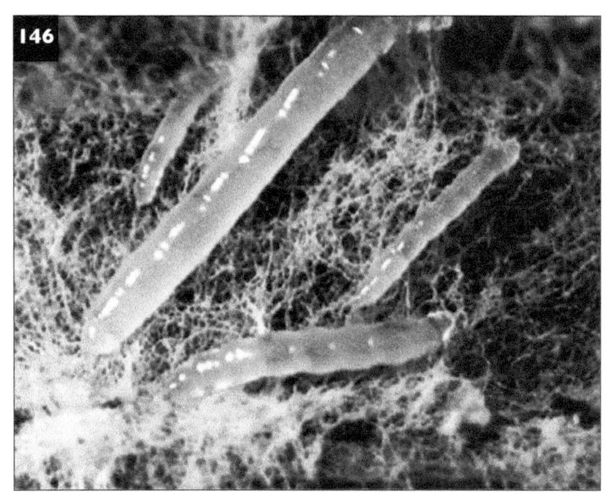

146 Cecid larvae of *Heteropeza pygmaea* showing size variation.

147 Cecid *Mycophila* mother larva bursting open to release daughter larvae.

Fertile females lay one or two eggs, a relatively unimportant means of reproduction, within the mushroom crop.

Epidemiology

As mushroom cecids can be readily found with fungal mycelium in a range of environments, such as decaying wood and rotten vegetation, it is possible that the flies from these natural habitats may give rise to farm infestations. Sometimes, initially, very small infestations may establish in peat; long-term survival in the wood of trays and other timbers used in cropping can also occur (*H. pygmaea*). After developing unnoticed through a number of generations, subsequent spread of larvae will give rise to further more serious outbreaks. The small sticky larvae are found on trays, on the hands, tools, equipment, shoes, and clothes of workers.

When present in large numbers, the larvae of *H. pygmaea* can swarm from the cropping-beds and fall on to the floor where they congregate in writhing heaps if the floor is dry. In addition to loss of marketable yield caused by their swarming on to mushrooms mostly in the latter half of the crop, *H. pygmaea* can also cause losses in total yield as a result of their ingestion of mushroom mycelium.

When undernourished, larvae of *H. pygmaea* develop fewer embryos but with thicker cuticles. In this condition, they can remain inert, but alive, for periods of up to 18 months. In contrast *M. speyeri* larvae, when starved, wander around in search of food for several weeks before they die.

Control

Larvae are easily spread about farms so it is vital to isolate infested houses using a phenolic disinfectant foot-dip (other disinfectants are not known to be effective for this pest). Similar disinfection must be used on tools and machinery that are likely to have contact with contaminated casing and dirty floors. Larvae swarming on the floor can also be killed with this disinfectant.

The use of separate protective overalls in every cropping house is advisable; effective cook-out of crops is very important. Dormant larvae, produced by *H. pygmaea*, are somewhat difficult to kill. It is important to ensure that spent compost is removed from farms. In addition, if the crop is in trays, or wood is used in the construction of mushroom beds, there is a strong probability that dormant larvae will survive crop termination treatment unless it is done thoroughly. Efficient phase II and cooking-out should kill all larvae: *H. pygmaea* is killed at 45°C, *Mycophila speyeri* at a slightly lower temperature. If cooking-out is not possible, thorough disinfection must be done.

Where trays of compost cannot be heat-treated, they must be very thoroughly cleaned after emptying and then cooked-out (the preferred option) or disinfected.

Casing materials should be stored hygienically before use: suspect materials should be discarded.

There is a danger that new wood, for trays and beds, will become contaminated if it is stored for long periods before being used.

Action points

- Effective cook-out will eliminate cecids.
- In the absence of heat-treatment, clean trays as thoroughly as possible with a disinfectant.
- Heat empty trays at 60°C long enough for the heat to penetrate the wood.
- To minimize spread, areas of infested crop should be isolated.
- Isolate infested houses with a phenolic disinfectant foot-dip.
- Observe strict hygiene throughout the farm (*see* Chapter 3).
- Make sure that casing ingredients are stored and mixed in clean areas.
- Do not stock-pile timber for new trays.

Non-paedogenetic cecids

Two species of non-paedogenetic cecid, *Lestremia cinerea* and *L. leucophaea,* are very occasional pests of mushrooms.

Symptoms

The dark orange (*L. cinerea)* or salmon-pink larvae (*L. leucophaea)* can be found feeding on developing mushrooms. *L. leucophaea* is found mostly at the base of stalks.

Description of pests

The bodies of adult female flies are 3–4 mm long, depending on the species, and their swollen abdomens packed with eggs are dull orange in colour. As their legs are very long and slender (**148**), both species bear a marked resemblance to mosquitoes. The larvae attain a length of about 4 mm when fully grown. Larvae of *L. cinerea* and *L. leucophaea* are dark orange and delicate salmon-pink respectively.

Epidemiology

Both species are common in the wild and there is evidence, especially with *L. cinerea,* that the females require natural daylight for egg laying. Their entire lives are spent on mushrooms rather than in compost.

Control

The exclusion of natural light is probably the single most important factor. It is likely that the chemicals used for phorid and sciarid control will kill adult *Lestremia* flies.

148 Cecid *Lestremia leucophaea,* a non-paedogenetic female. Adults have a leg span of about 12 mm.

Other flies

Sphaeroceridae (sphaerocerids)

Some species of this family, which superficially resemble phorids, occur in manure during phase I composting (**149**). The commonest species in mushroom houses is *Pullimosina heteroneura,* which can be differentiated from mushroom phorids by the red colour of its eyes and its unique wing venation (**150**). It breeds in compost which is unsuitable for mushroom mycelium and where intense bacterial action is taking place.

The adults of *Pullimosina* are carriers of mites and pathogenic fungi.

The sporadic occurrence of this pest has not been fully explained, but it is possible that large local populations associated with manure or other forms of rotting organic debris may endanger crops. There is no specific information available on control, but recent observations suggest that the use of parasitic nematodes may be of benefit.

Drosophilidae (fruit flies)

Drosophilid flies (*Drosophila funebris* has been found in mushrooms) are normally associated with decomposing vegetable matter, especially fruit. However, mushrooms can occasionally be tunnelled by larvae which can be numerous; affected mushrooms can be reduced to liquid. Eggs are laid just below the surface of cap tissue causing a brown spot which, when it expands, resembles an early stage of Bacterial blotch.

As these insects are opportunistic pests, specific control measures are not justified. Scrupulous hygiene, which should be the norm, would avoid accumulation of their favoured substrate (organic debris).

149 Sphaerocerid, adult fly of *Pullimosina heteroneura*.

150 Sphaerocerid; typical wing venation.

Scatopsidae (scatopsids)

The larvae of these flies, which are usually found in manure or decaying vegetable matter, can breed in mushroom compost while bacterial decomposition is occurring. However, they are not thought to damage mushroom crops, albeit adult flies are sometimes caught on sticky traps within mushroom houses.

Wood gnat (Anispodidae)

Wood gnats have been found in a number of different locations in eastern USA. The distinctive adult fly is 11–15 mm long with mottled clear and light brown wings. The antennae are dagger-like, projecting forward from the head; they are attracted to lights. Eggs may be laid on mushroom stalks and maggots can burrow into the mushroom and feed inside stalks and caps. No part of their life cycle is spent in casing or compost.

These flies are probably susceptible to the commonly used insecticides, and should also be controlled by the removal of mushroom trash.

Moth flies (Psychodidae)

Moth flies or drain flies are common household nuisance pests. They breed in water film in drains or in water collection basins. The adults are small (about 6 mm) and hairy, resembling a tiny moth. At rest, the wings of the adult are held over its back in the shape of a peaked roof. Although regarded as a nuisance in mushroom growing rooms, they might be significant as vectors of mushroom pathogens, mites, nematodes, and of food-borne human bacterial pathogens found in drains.

The adults are likely to be susceptible to the insecticides that affect the adults of other flies; their immatures should be eliminated by cleaning drains and water collection basins at frequent intervals.

Mushroom mites

Mites are very common but are not often noticed until they are present in large numbers. Compost ingredients and phase I mixes provide a suitable habitat for many species, and large numbers can occur in the outer layers of phase I stacks. Most are predatory, feeding mainly on other species of mite as well as nematodes. Some feed solely on weed moulds, and some on bacteria. They are all heat-sensitive and therefore are unlikely to survive effective phase II composting, but because of variations in composting temperatures, inevitably some survive.

Tarsonemid and pygmy mites feed on mushroom mycelium and, together with red-pepper mites, are the most important mite pests of mushrooms. Most of the others do not directly affect the crop. However they are effective carriers of fungal spores and could be a means of distributing pathogens of mushrooms from crop to crop.

In large numbers, mites can annoy pickers: some allergic reactions and skin irritation have been perceived to be associated with their presence.

Tarsonemid mites

In some countries, tarsonemid mites (mainly *Tarsonemus myceliophagus*) are the most important pests of mushrooms but have steeply declined in importance with improvements in compost preparation and crop termination.

Symptoms

Mites cause a reddish-brown discolouration and rounding of the base of affected mushroom stalks, sometimes entirely severing the basal attachments of mushrooms (**151**). Very rarely, the whole mushroom may be discoloured. As mites reproduce slowly, damage is generally not seen until the third flush. However, where an infestation occurs soon after spawning, large populations may develop, with subsequent increases in amounts of damage.

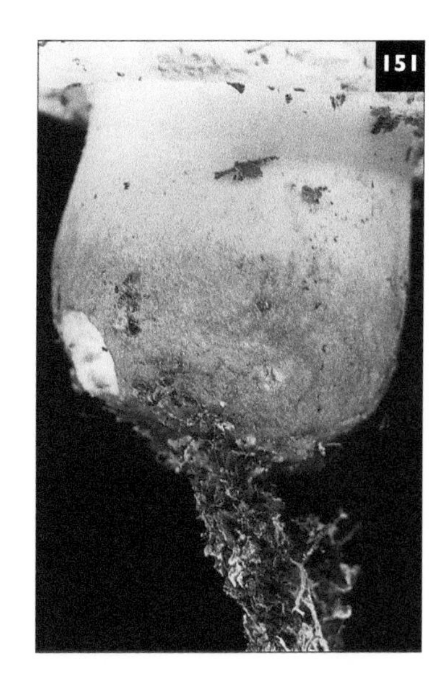

151 Tarsonemid mites causing typical damage to the base of the stalk, which is only partially attached to the casing.

Description of pest

T. myceliophagus is an oval pale brown shiny mite which is so minute (0.18 mm) that it is virtually invisible to the naked eye. Compared with most mites and insects, it has a fairly slow rate of increase, producing (on average) one egg per day over a period of 2–3 weeks. At 16–24°C the life cycle from egg to adult is completed in 11–12 days. At 16°C the adult females can survive for 2 months, whereas at 24°C the average length of life is 15 days, although a few may remain alive for a month. The mites feed on the mycelium of mushrooms and of various weed moulds (*Chaetomium, Trichoderma,* and *Penicillium* spp).

Within the crop, largest populations of mites are located in casing. If this is allowed to dry out, the female mites stand erect on the casing surface, and jump about 5 cm into the air. Air movement may then transport them to a more favourable environment.

Laboratory tests show that adult mites are killed by exposure at 49°C for 20 minutes in a saturated atmosphere, but living mites have been recovered from compost cooked-out at 60°C, possibly because they retreat to the shelter of cool crevices if treatment has not been uniform.

Control

The decline in the importance of this pest in recent years has coincided with improvements in composting and cook-out. An effective phase II and a thorough and efficient cook-out should be all that is necessary to achieve control.

Action points

- At the end of cropping, ensure that an effective cook-out temperature is reached throughout the growing medium, trays and/or shelves.
- Observe strict hygiene throughout the farm (*see* Chapter 3).
- Ensure that all machinery and rooms involved with spawning and spawn-running are thoroughly cleaned.
- Make sure all spent compost is removed from the farm.

Pygmy mites

The true Pygmy mite (*Microdispus lambi* syn. *Brennandania lambi*) is quite distinct from the red pepper mite, not only taxonomically, but also in its biology. It occurs in China and Eastern Australia, where it is rated as probably the most serious pest of mushrooms. Since 1997, it has been a serious pest in Spain but so far it has not been reported in Northern Europe or in North America.

There is some confusion with the names of different mites that affect mushrooms, and in North America the red pepper mite (*Siteroptes mesembrinae* syn. *Pygmephorus mesembrinae*) is sometimes referred to as the pygmy mite.

Symptoms

Pygmy mites swarm in vast numbers generally towards the end of the crop. When severe, the whole of a third flush can be covered by swarming mites (152). In extreme conditions, pygmy mites swarm

152 Pygmy mites swarming on the casing of a crop. By kind permission of Dr Alan Clift.

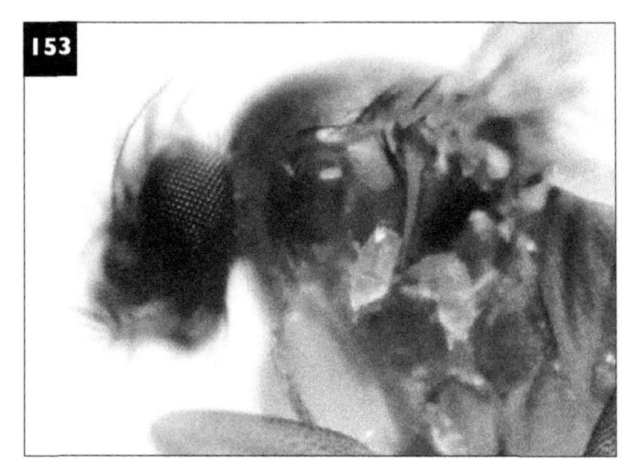

153 Phorid fly with pygmy mites attached. By kind permission of Dr Alan Clift.

much earlier, even at the first flush stage. They feed on mushroom mycelium and as a result can cause considerable crop loss.

Description of pest

Pygmy mites are similar to red pepper mites in appearance, being brownish in colour, but are smaller (0.18 mm in Australia although reported at 0.25 mm in Spain. Red pepper mites are 0.20–0.25 mm). Eggs are laid in compost and hatch into larvae which become nymphs before being adults, all in a period of 10 days. The adults live for a period of 7 days during which time they can move from crop to crop.

Epidemiology

There is generally a period of about 5 days when the adult mites are transported by flies. Their rate of reproduction is very quick once introduced into a crop where they feed on actively growing mycelium. At the end of a crop there can be very large populations of both flies and mites. There is no association with the weed moulds found in mushroom crops, nor have they been found in any of the compost ingredients. For their distribution adult mites are entirely dependent on flies, in particular phorids but also sciarids – such an association is known as 'phoresy'. This term implies the attachment of one or more of the development stages of one organism to a carrier organism which is enhancing the dispersal of the less mobile partner. This phoretic behaviour in which mites associate with flies is characteristic of pygmy and other mites: the predatory mites. In order to be transported, the mites adopt a characteristic perching stance on the crop surface where the females rear up and grasp any passing object, including flies. Phorid flies have been observed carrying as many as 60 mites per fly, but much smaller numbers are generally found on flies entering a new crop, the principal method of transfer (**153**).

Control

Control is achieved by the exclusion of the mite vectors (*see* Control of sciarids and phorids, pp. 145 and 149).

Red pepper mites

Commonest mites seen during mushroom cropping are red pepper mites. *Siteroptes mesembrinae* (syn. *Pygmephorus mesembrinae*) is the most common species, but others have been recorded in the UK and in the USA. Red pepper mites are not normally regarded as primary pests.

Symptoms

These mites often swarm in vast numbers on the surface of casing and on mushrooms, giving them a reddish-brown colour (**154, 155**), hence their name. They feed only on weed moulds, particularly the spores of *Trichoderma* spp.; they do not feed on mushroom mycelium. Their presence in pre-packs and in boxes, where they migrate to the surface, can result in crop loss due to spoilage.

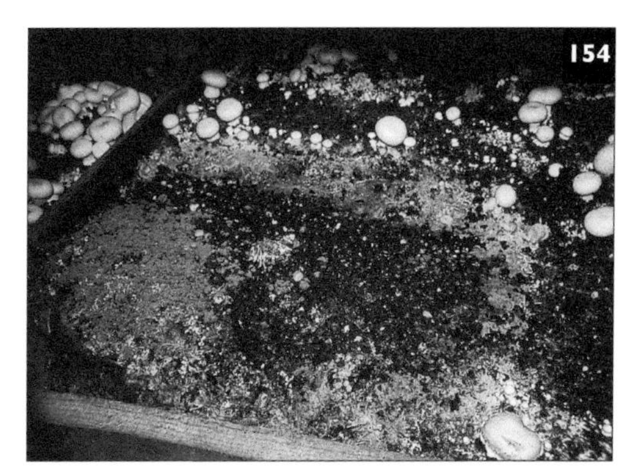

154 Red pepper mites (*Siteroptes* spp.); a large cluster on the casing surface together with *Trichoderma aggressivum*.

155 Red pepper mites; mushrooms covered with jostling swarms.

Description of pest

These mites are tiny (0.20–0.25 mm), have a flattened wedge-shaped appearance and are yellowish-brown in colour with a central whitish internal band. *Siteroptes mesembrinae* is dependent upon *Trichoderma* for breeding. It is capable of rapid rates of increase as females can lay up to 150 eggs over a period of 5 days. These hatch and the life cycle from egg to adult is completed in 4–5 days at 20–25°C.

Red pepper mites often indicate the presence of *Trichoderma* in the compost. They are known to carry the spores of *Trichoderma* spp. (*see* p. 86).

Control

Red pepper mites are secondary, and it is the elimination of the primary problem that prevents their development. In recent years, they have been closely associated with Trichoderma compost mould. Control of red pepper mites is therefore dependent upon the rapid establishment of the crop, in particular immediately after spawning.

Because of their ability to spread *T. aggressivum* (*see* pp. 86 and 88), it is sometimes considered desirable to treat isolated areas of red pepper mite infestation. Although both salt and flame guns have been used with some success, it should be emphasized that surface treatments will not correct the basic problem, nor kill all the mites.

Action points

- Make sure that composting and phase II are efficient (*see* Chapter 3).
- Control Trichoderma compost mould (*see* p. 88).
- Observe strict hygiene, particularly at spawning.
- Apply spot treatments particularly where there is a Trichoderma problem (*see* p. 88).

Predatory mites

There are three species of predatory mesostigmatid mites (often called gamasid mites) which are frequently encountered in mushroom houses. They feed on different developmental stages of most mushroom pests and may be beneficial, although their potential as biological control agents has not been exploited. Their presence is indicative of large numbers of their prey species, and in this respect they are indicators. Like pygmy mites, they establish a phoretic relationship with a range of mushroom flies, most notably sciarids; they do not associate with phorids.

Symptoms

Predatory mites do not trigger symptoms on mushroom crops. They can frequently be seen running over casing, mushrooms, trays, and the hands of pickers, in search of prey species. They are said to be a cause of skin irritation to pickers.

Description of mites

They are pale orange to dark red in colour, and are easily distinguished from pest species by their larger size and/or their greater speed of movement.

Parasitus fimetorum has a phoretic association with dung beetles, and not flies. It is about 1 mm in length and is often seen running on the surface of casing and on the sides of containers. It only attacks mobile prey such as nematodes, insect larvae, and other mites. Though recorded, it is not important in mushroom crops.

Digamasellus fallax has a phoretic association with sciarids. It is about 0.5 mm long and feeds on rhabditid nematodes. It is likely that it would also eat mycophagous nematodes.

Arctoseius cetratus has a phoretic association with sciarids. It is about 0.4 mm long and feeds on a wide range of small creatures, including nematodes and fly larvae.

The predatory mites range in length from 0.4 mm (*A. cetratus*) to 1.0 mm (*P. fimetorum*). Both species are polyphagous and eat a wide range of prey, including insect eggs and larvae, mites, and nematodes.

Epidemiology

D. fallax and *A. cetratus* can be introduced into mushroom crops by flies, especially sciarids, as they are able to attach themselves to the abdomens of their more mobile hosts.

Control

No measures should be taken to control these mites. Rather, the crop should be examined for sources of food, and on these findings, appropriate measures should be taken against the actual food pest that is present.

Action point

- Identify the pest or pests on which predatory mites are feeding and take appropriate control measures.

Other mites associated with mushroom crops

This group of mites includes *Tyrophagus* spp., *Caloglyphus* spp., *Histiostoma* sp., and *Linopodes antennaepes*. They are not known to feed on mushroom mycelium, although, in the past, some have been associated with damage to crops. Some may feed on the mycelium of moulds but as all feed on decaying tissue they are said to be saprophagous. Small numbers of these mites may be found on the surface of the casing.

Symptoms

In the past, some of these mites (especially the *Tyrophagus* spp.) have been associated with small pits in the caps of mushrooms, as well as browning of the stalks which then become colonized by bacteria and decay (*see* Bacterial pit, p. 102). It is not clear however, whether the mites are causing the damage, or are merely exacerbating an existing bacterial attack. Tyrophagus mites may feed on mycelium of other fungi, and have been shown to be able to spread the spores of *Verticillium* spp. The other mite species feed, perhaps exclusively, on decaying mushrooms. *Linopodes antennaepes* has often been found in association with browning of the base of the stalks of mushrooms caused by tarsonemids.

Description of pests

These are soft translucent white mites whose bodies carry long hairs (except *Histiostoma* sp.). *L. antennaepes* is light yellowish-brown in colour, with long front legs which are more than twice the length of the body (commonly known as the long-legged mite). They are very fast movers on the bed surface, and although once common are now comparatively rare. Generally all of these mites are larger (0.3–0.6 mm) than tarsonemids and red pepper mites, and apart from the long-legged mites, are slower in their movement. They are normally only found in areas of decomposition into which mushroom mycelium cannot grow. Consequently they are often associated with saprophagous nematodes (*see* p. 163).

Epidemiology

Tyrophagus spp. have a phoretic association with flies, and are usually carried to compost on the bodies of flies; they cling by means of suckers. The migratory stage is usually produced when mites become overcrowded.

Control

With efficient composting and phase II, predatory mites are unlikely to be a nuisance. However, fungal and bacterial contaminants are commonly found at the end of the cropping and, as these mites breed readily on such substrates, an efficient cook-out and safe disposal of the spent compost is essential. Good general hygiene, especially in the clearance and elimination of organic debris, is vital.

Action points

- At the end of cropping ensure that an effective cook-out temperature is reached (*see* Chapter 3).
- Eliminate the vector species by controlling flies in particular (*see* pp. 145 and 149).
- Observe strict hygiene throughout the farm (*see* Chapter 3).

Nematodes

Many different nematodes (eelworms) are found in mushroom compost and casing. Their effects on mushroom crops depend upon the species, the size of populations, and especially their feeding habits. Nematodes in mushroom beds can be classified into three groups. (i) There are those that feed by sucking up liquid together with finely particulate organic matter and bacteria – these are the saprophagous forms, and in mushroom beds they are predominantly in the group referred to as rhabditids (**156**); (ii) there are the attendant nematodes that feed on rhabditids – they form the predaceous group; and (iii) there are those nematodes that feed on fungal mycelium (mycophagous), the tylenchids – they are damaging mycelial feeders and adversely affect mushroom crops.

In recent years, with improvements to composting and casing materials, mycophagous nematodes have become less common, although saprophagous types remain but are rarely associated with crop losses.

Description of pests

Nematodes are small, colourless, and up to 1 mm long. They swim in surface films of water in or on compost and casing. Without water they cannot move and do not reproduce. The mycophagous nematodes are distinguished from the saprophagous types by the possession of a spear at the head end which is used to penetrate fungal mycelium and suck out the contents (**157**).

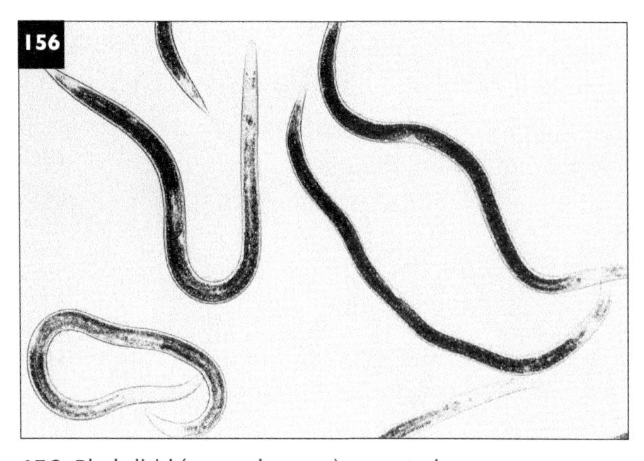

156 Rhabditid (saprophagous) nematodes.

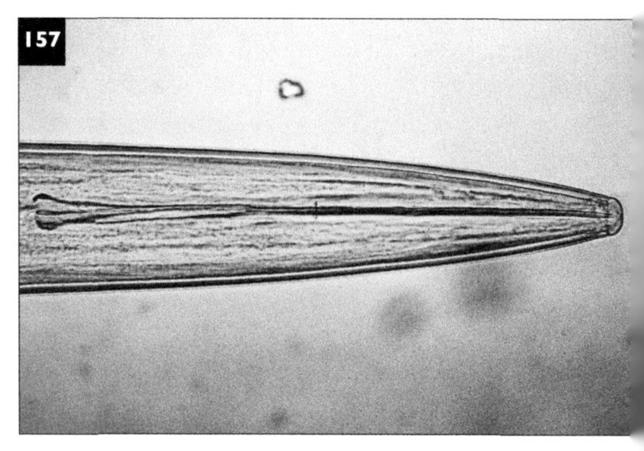

157 Parasitic nematode; the head end showing the feeding spear. By kind permission of Dr Mike Copland.

Saprophagous nematodes

The saprophagous nematodes, mostly rhabditids, form the largest group of nematodes to be found in mushroom beds. Rhabditids can increase very rapidly (50–100 fold per week). They are not normally regarded as primary pests because they feed on decaying organic material. Their presence within a crop, therefore, is an indication of poor composting particularly in phase II. An inefficient pasteurization will result in compost that is unfavourable for the growth of mushroom mycelium, and may also allow nematodes to survive. In bulk phase II, the compost at the ends of a tunnel is most likely to be under-heated, and it is in this compost that nematodes may be found. Survivors can lead to abnormally large nematode populations in compost at spawning. Any part of the phase II treatment that does not reach the maximum temperature is not only likely to contain surviving nematodes, but will also be poor compost. The presence of nematodes in casing can also indicate ineffective hygiene, although some peats have been known to have high nematode populations.

Predators, including mites, fungi, and predaceous nematodes, can attack rhabditids. One fungus, *Rhopalomyces elegans*, parasitizes the eggs of rhabditids. This fungus is worldwide in its distribution, and has been shown to be a host of *Verticillium psalliotae* and *Mycogone perniciosa*. Whether this association has significance as a source of mushroom pathogens is unknown.

Large numbers of saprophagous nematodes have sometimes been associated with crop losses. It is not clear whether these losses are the result of direct damage to mushroom mycelium, or are the result of a secondary effect.

Symptoms

Nematodes swarm in vast numbers on the surface of casing, where they can be seen glistening in the light (**158**). Bad harvesting practices, such as leaving cut stalks and stumps on the beds, can encourage the development of these nematodes which feed on decaying mushroom tissue (**159**).

158 Swarm of rhabditid nematodes on the surface of a mushroom bed. These are best observed with a light held horizontal but close to the casing surface. The aggregated nematodes can then be seen moving.

159 Rhabditid nematodes feeding on a decaying mushroom. The white spikes on the remains of a mushroom stalk are large numbers of aggregated rhabditid nematodes. These spikes can be seen moving, particularly if the air above them moves.

Mycophagous nematodes

Ditylenchus myceliophagus and *Aphelenchoides composticola,* two species of mycophagous nematodes, are primary pests of mushrooms; they feed exclusively on fungi and can destroy mushroom mycelium. Their rate of increase is very rapid and may be as much as 25,000 times in a week. In adverse conditions they can survive for up to 6 weeks without food. They are now very infrequently seen in commercial crops.

Symptoms

If contamination occurs early, numbers of nematodes can become sufficient to destroy mycelium. In infested compost, fine mycelium disappears leaving only the larger strands which give the compost a stringy appearance. If cropping has ceased, little or no mycelium can be found, and the compost becomes dark and sodden, with a distinctive pungent smell. The patches can increase in size throughout cropping. Severely affected areas may show a grey mould growth of nematode-trapping fungi that are sometimes associated with nematode infestations (see *Arthrobotrys* spp., p. 133). When their food source has been exhausted, nematodes migrate and congregate in millions, forming curds in the compost, on the surface of casing and growing containers. As they dry, the curds of nematodes curl and may eventually break up and be distributed to other areas. At this stage they are readily distributed on hands and by insects.

The extent of yield loss is dependent on the timing and severity of the original infestation. An infestation at spawning can cause spawn-run to fail, and render growing completely uneconomic. A later infestation, towards the end of cropping, would cause only a slight loss in yield and would probably go unnoticed by growers.

A. composticola feeds on many fungi, including mushroom mycelium. It breeds very rapidly, and in severe infestations these nematodes tend to adhere to each other, forming whitish clumps.

Control

Large populations of saprophagous nematodes generally indicate problems in phase II composting. In addition, they have been known to occur in some peat samples. An on-farm investigation of phase II temperatures and a check of casing hygiene are required in these circumstances.

When an attack by mycophagous nematodes has been initiated, little can be done apart from destroying the compost from affected areas plus some of the surrounding, apparently healthy compost. *D. myceliophagus* can withstand drying for up to 3 years, and this makes the safe disposal of all spent compost very important, as dry infested debris is a potential source of trouble.

Cooking-out at the end of the crop is by far the best method of nematode control. However, nematodes are extremely difficult to eradicate, and can be carried over from one crop to another lodged in cracks in boards and trays. A uniform temperature of 55–60°C must be achieved throughout the compost, and in trays and bed boards.

Action points

- Identify the type of nematode present.
- Ensure that temperatures during phase II reach 60°C and are uniform (*see* p. 57).
- At the end of cropping, ensure that an effective cook-out temperature is reached throughout the growing medium, trays, and shelves (*see* p. 46).
- Make sure the casing ingredients are stored and mixed in clean areas (*see* p. 14).
- Observe strict hygiene throughout the farm.
- Make sure that all the spent compost is taken away from the farm.

Minor pests

There are several other invertebrates that occasionally reach pest status. They tend to be associated with bad growing conditions or unconventional growing systems; in nature they are mostly either saprophagous or ground-dwelling in nature.

Collembola (springtails)

If mushroom beds are on a bare-soil floor, springtails may reach pest status. They favour damp conditions and an abundance of decaying vegetable matter. Mushrooms grown in glasshouses as catch crops are especially vulnerable, as enormous numbers of springtails may develop and cover the surface of the mushroom beds. Springtails can also be found in large numbers in the surface layers of manure stacks.

The commonest species is *Archorutes armatus*, which is slate-blue in colour. It can feed on both mycelium and sporophores, causing minute open pits to develop on stems and caps of mushrooms. From these pits, dry branched tunnels can be formed.

Springtails can be eliminated by efficient pasteurization. The floors of cropping houses should be kept free of organic debris and, if possible, crops should be raised off the floor.

Diplopoda (millipedes)

Millipedes may occur when crops are grown directly on soil; they eat holes into the bases of mushroom stalks. However, apart from keeping the floors of cropping houses clear of all organic debris, no other control measures are justified.

Gastropoda (slugs)

Slugs may cause some trouble by eating large cavities into caps and stalks of mushrooms, but this rarely occurs because most structures effectively prevent their access. Control measures are rarely justified but metaldehyde baits could be used, but always at a distance from crops.

CHAPTER 9

Abiotic Disorders

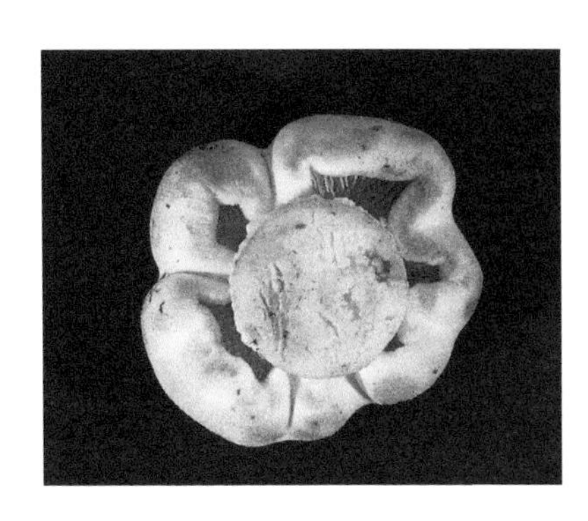

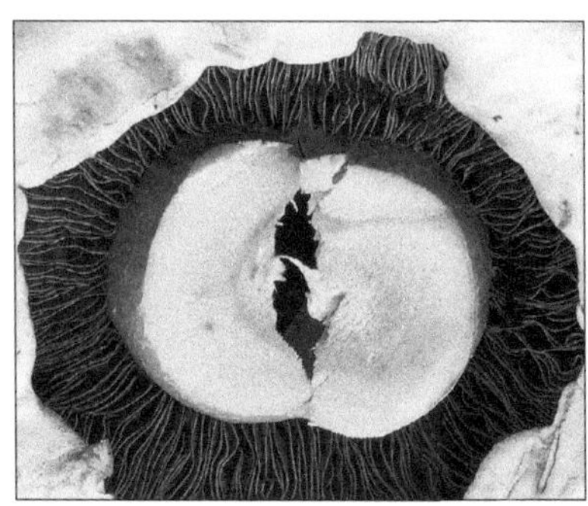

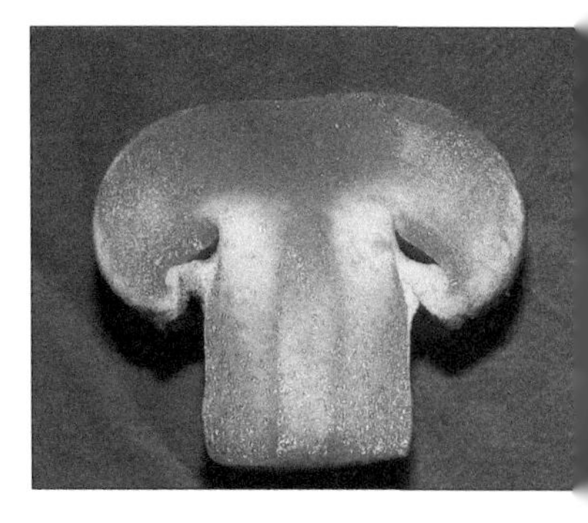

Introduction

Abiotic disorders are those not known to be caused by a pest or pathogen (biotic cause). Many result in symptoms very similar to those with a biotic cause. This is not surprising as there is a limit to the variety of symptoms that mushrooms can produce. In the future, it is possible that a biotic cause may be found for some although it is likely that most have none. It is helpful for a mushroom producer to be familiar with these problems in order to be able to differentiate them from the others. In general, abiotic disorders do not constitute a regular cause of crop loss but they can sometimes be very significant. They are generally erratic and unpredictable in occurrence. Many occur once only. They are extremely difficult to reproduce in experiments, even when the conditions in which they were found are simulated as accurately as possible. The disorders are described in a crop development sequence, starting with precropping, then cropping (with whole mushroom symptoms, stem and cap disorders grouped), and finally harvesting and post-harvesting.

Pinning disorders

There is a range of pinning disorders with more or less identifiable causes. These problems are now less common, but can be serious as they affect the management of harvesting, the quality of the crop, and often the total yield.

Stroma

When the mycelium grows through the casing and appears on the surface to form a mat, the symptom is called 'Stroma' or Overlay. The name Stroma in this book is used where the mycelial growth is very limited in distribution, forming small patches which are rather fluffy in appearance and slightly brown or off-white in colour. This type of mycelial growth, also referred to as Sectoring, has been considered to be the result of a mutation or change in the strain. With some strains it is seen on the surface of the compost at the completion of the spawn-run and before the crop is cased. It is not generally a cause for concern.

Overlay

Overlay is where the mycelium is not abnormal in appearance but forms a white capping on the casing surface. It can be impervious to water, preventing cropping. The cause is an interaction of environmental factors, including the casing temperature, carbon dioxide concentration, the amount of evaporation from the casing surface, excessive casing inoculum, or the amount of water held in the casing. Casing depth can be important, as chronic overlay often occurs when the casing is shallow and the mycelium reaches the surface quickly. Variable compost temperatures can affect the casing temperature, which can cause more rapid mycelial growth in the warmer areas. Cooler parts of the casing are colonized more slowly. Uneven compost temperature can result from uneven fill with the areas with most compost becoming active and warmer more quickly.

The wetness of the casing and the amount of evaporation affect the casing temperature. Often a change to a more open, better draining casing will reduce overlay. A persistent problem is most likely to result from poor environmental control. Apart from rectifying the environmental conditions, the only recourse, once the problem has occurred, is to ruffle (lightly rake) affected areas as soon as they are seen, in order to slow down the mycelial growth in those areas.

Mass pinning

This disorder may be recognized when unusually large numbers of pin-heads are formed. Such crops produce many small mushrooms, or none if the pin-heads develop into a layer of partially differentiated mushroom tissue, which precludes the production of mushrooms. This extreme condition has effects similar to those of severe Overlay.

Mass pinning occurs when, in an environment that favours primordia initiation, the temperature and atmospheric carbon dioxide concentration drop very rapidly over a 24-hour period. Extending the period of temperature decline to 3 or more days, with a gradual drop in the carbon dioxide concentration, reduces the pin number. Very high concentrations of carbon dioxide inhibit pin-head formation. These factors, together with watering, are used to regulate numbers of mushrooms produced.

If the problem persists, the environment must be examined to identify other aberrations that may have occurred. Sometimes it is made worse by the external environmental conditions, especially on farms where air conditioning equipment is not satisfactory. Other factors which can be important are casing type, shallow or dry casing, and late airing.

Pin death

Pin-heads may die even though the casing is well colonized, creating non-cropping patches (**160**). The symptom often follows over-pinning and is a natural phenomenon. It becomes a problem if more than the normal numbers of pins die. Poor cultural conditions, such as pools of water after watering, and excessively high compost temperatures, can result in pin death. The presence of an impervious mycelial layer in the casing, generated early in the life of the crop, can result in poor drainage of the compost and water logging, and subsequent pin death.

Affected crops should also be checked carefully for sciarid larvae (*see* p. 143) which may also be a contributory factor.

Clusters or clumping

Some crops, particularly in Holland, have shown excessive clustering. Clusters are fused clumps of mushrooms with deformed shapes, often formed on the highest points of the casing, or on exposed compost (**161**). They tend to be most conspicuous in first flushes. The growth rate of mushroom mycelium made from such clusters is greatly reduced (reported to be about 20% of that of mycelium from normal mushrooms). Spawn made from such affected cultures fails to grow at all in mushroom compost. It appears that the only way a cluster culture can thrive is by parasitism of a healthy culture. In an experiment, 0.2% cluster culture mixed with 99.8% healthy spawn resulted in 40% of all subcultures of the clustering type, taken 3 weeks after mixing (*see* Production of brown mushrooms in MVXD, p. 123). It appears that parasitism of healthy cultures has an infectious characteristic in which cluster mycelium is able to convert normal mycelium to the cluster type. Research results indicate that Clusters may be associated with a transmissible element, presumed to be of nucleic acid origin. It is believed that this element may be active transposons (messenger genetic material). Molecular analysis of affected spawn shows the presence of at least six new retrotransposons. Preliminary data suggest that there is a possible association between altered transposon positions in the *A. bisporus* genome, and the shift from healthy to the cluster state. Tests of clusters for virus diseases, both La France disease and MVXD disease, were negative.

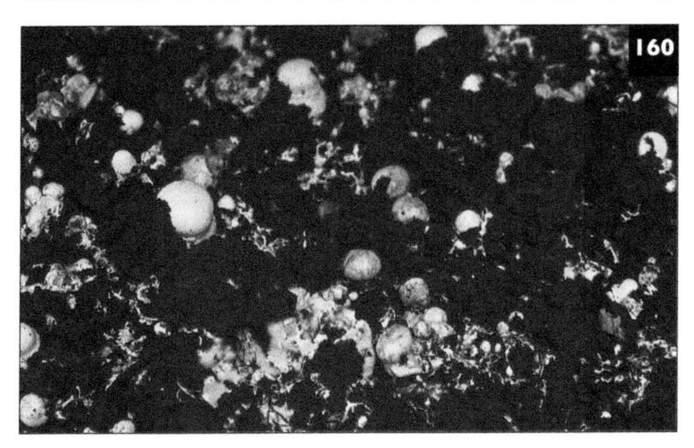

160 Brown and dead small mushrooms in a third flush of mushrooms.

161 A small Cluster of mushrooms in the early stages of development of the first flush. Note the mass of mushroom tissue with some mushrooms developing on top of each other.

Whole mushroom distortion

Mis-shapes

In one of the most commonly seen forms of distortion, mushrooms appear to be incompletely differentiated. The symptoms are most commonly associated with first flushes, particularly of autumn crops, and they are sometimes strain-linked. Mushrooms which fail to develop the normal shape can range from rather knobbly lumps to recognizable mushrooms with rather grotesque mis-shapen caps. Fusing of individual mushrooms can also occur (**162**). Early first flush mushrooms are frequently partially formed or are fused multiples. This is probably the result of suboptimal conditions at the time of fruit body initiation. If these symptoms occur, the only recourse is to examine the past environmental conditions for some aberration.

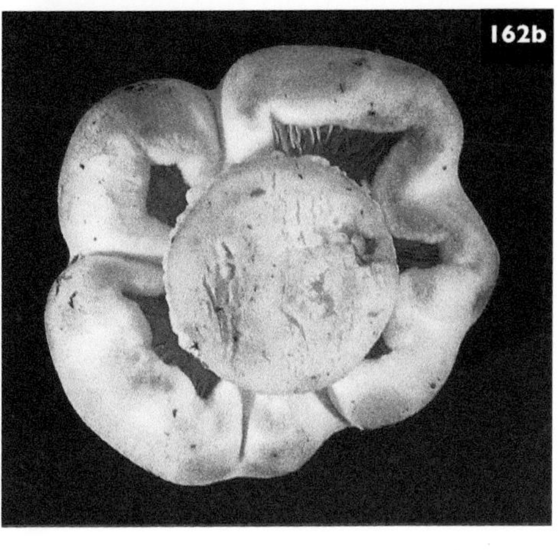

162 Two different sorts of mis-shapen mushrooms. (**a**) Two mushrooms have fused together at their bases and their caps and in (**b**) the edge of the mushroom has developed in a corrugated way. The cause of both of these symptoms is unknown, but such mushrooms are often more common in the first flush.

163 An unusual distortion recently seen in which the developing mushrooms have the shape of hot-air balloons.

Balloons

On at least two quite disparate sites in 2004, mushrooms were produced which were reminiscent of hot-air balloons in shape (**163**), with no differentiation into stipe, cap or gills. Their numbers within affected crops were small, but it was reported from both sites that there was a reduction in yield and quality.

No explanation for this unusual phenomenon is yet available. Virus tests on mushrooms from the two sources proved negative. Cultures from 'Balloons', when cropped, failed to reproduce symptoms.

Carbon dioxide damage

The main symptom of a high concentration of carbon dioxide is elongation of stalks: affected mushrooms also have small caps. In extreme cases they are described, as 'drumstick mushrooms' (*see also* Virus disease, p. 115 and **109**). Sometimes crops are grown at higher than normal carbon dioxide concentrations, either to elongate the stalks slightly to make mechanical picking easier, or to regulate the numbers of mushrooms that develop.

Mummy-like symptoms

A disorder with dry mummified mushrooms similar to those of Mummy disease (*see* p. 100), referred to as False mummy or Crypto-mummy, was not uncommon some years ago but is now rare. Its occurrence was quite widespread and was distinguished from Mummy disease by the ability of the crop to recover.

The cause of the disorder is unknown, but has been associated with chronic over-watering. A viral cause of this problem cannot be ruled out, but crop recovery coincidental with a reduction in water use suggests an abiotic cause.

Mushroom stems

Hollow stems and split stalks

The affected mushrooms may appear to be normal on the bed, but when they are cut for market they are found to have partially hollow stems, often with a circular cavity surrounding a solid core (**164, 165**). The cavity may extend from the base of the stalk to the cap, or may be smaller and incomplete. In extreme cases the cut surface of the hollow stalk may split and curl backward (**166**). The stalk may also split at cutting, or even before, to give a Chinese lantern effect, sometimes referred to as Saggy socks (**167**). The vertical splits may be accompanied by horizontal ones, which enable strips of stem to curl upwards, downwards, or both.

These symptoms are associated with problems involving water, and may occur following a period of excessive water uptake followed by a period of rapid evaporation. The water content of casing and compost, and atmospheric humidity, have all been shown to have an effect.

Environmental conditions, in particular evaporation, casing type, and compost water content, should be examined if these problems persist.

164 Hollow stem not seen until mushrooms are harvested. The tissue around the centre of the stalk is dry and stringy.

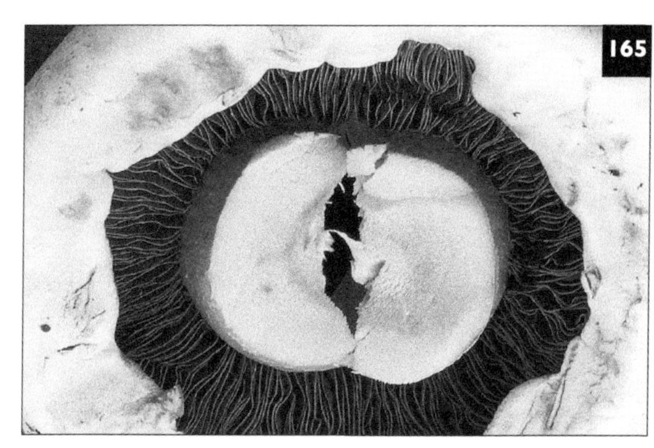

165 A hollow stem which has split longitudinally. Often hollow and split stems become colonized by bacterial soft-rotting organisms, producing a stem decay symptom.

166 The centre of the stalk of this mushroom is hollow and the sides of the stalk have split and curled. The whole mushroom is very wet.

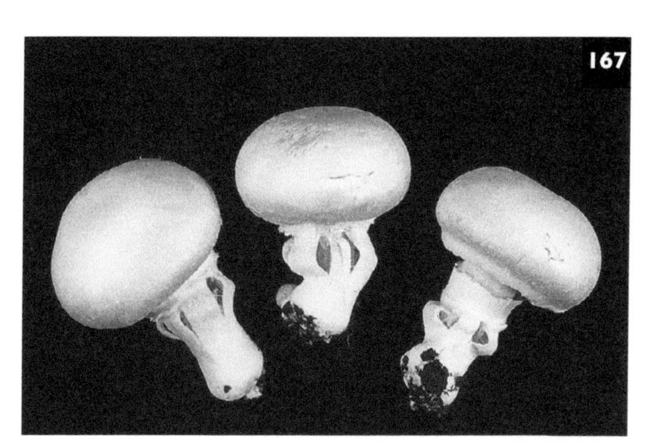

167 The stalks of these mushrooms have split before harvest and sag to give a Chinese lantern or Saggy-socks symptom.

Swollen stems

Mushroom stems have been seen that have large even swellings, so that, instead of the roughly cylindrical shape of a normal stem, there is a concentric bulge anywhere on the stem (**168**). In developing mushrooms the base of the stalk may be considerably enlarged in comparison with the cap, and these have been referred to as Cottage loaves. With further development of the cap, this symptom generally disappears.

Swollen stems are unusual and their cause is not known.

168 Stems may bulge at their bases to give cottage loaf symptoms or the bulges may occur higher up on the stem.

Mushroom caps

Early openers

Affected mushrooms may open before their maximum size is reached (**169**). Such symptoms are usually seen in the later flushes. The affected mushrooms are of poor quality and are often unmarketable. The most likely cause, other than a pathogen (*see* pp. 114 and 128), is the restriction of water entry into the mushroom at a critical stage in its enlargement. Sometimes gross over-pinning can have this result. Factors affecting water uptake should be investigated if the problem continues and the presence of a pathogen has been ruled out.

Hard gill

Open mushrooms have pale-coloured gills which are very stunted or even non-existent (**170**). If the cap is examined by breaking, it is often unusually thick, and there is little or no gill tissue. Environmental factors, in particular periods of excessive evaporation, are thought to be at least partly responsible for the symptoms. There is also a tendency for some strains to be more frequently affected, although Hard gill can occur in all of them. Similar gill symptoms are sometimes associated with virus diseases (*see* p. 114).

Rosecomb

This is a specific distortion of mushroom caps where pink or rose-coloured gill tissue, often with a porous appearance, develops on the surface or at the upper edge of the mushroom cap. It may be in warts, or in vertical structures often referred to as Cocks' combs. Mushrooms so affected are grotesque and unsaleable (**171**). The cause has long been attributed to contamination by hydrocarbons, phenol, and other compounds. Diesel oil and the exhaust from diesel or petrol engines, and some of the ingredients of creosote, are thought to cause this type of distortion. Contaminated casing is often a primary cause. Heavy overdosing with certain pesticides, particularly oil emulsions, has been known to be a cause.

The source of the distorting material must be identified.

Clefts and craters

Deep clefts in the surface of caps, which sometimes result in splitting, are a symptom that often occurs, together with more general distortion. The cap surface may have a distinct depressed area (**172**). Distortion of this kind can be seen on all strains but rough-white strains are particularly prone to the deep-cleft symptom. Between-flush mushrooms are commonly affected by these symptoms. This suggests that incorrect environmental conditions at the time of the expansion of the developing mushroom may be responsible. Fortunately, the problem tends to be transitory, and usually does not affect a large proportion of a crop.

169 Sometimes mushrooms open well before they have reached their full size. Early openers can result from disease problems such as virus or can reflect a lack of water in the compost.

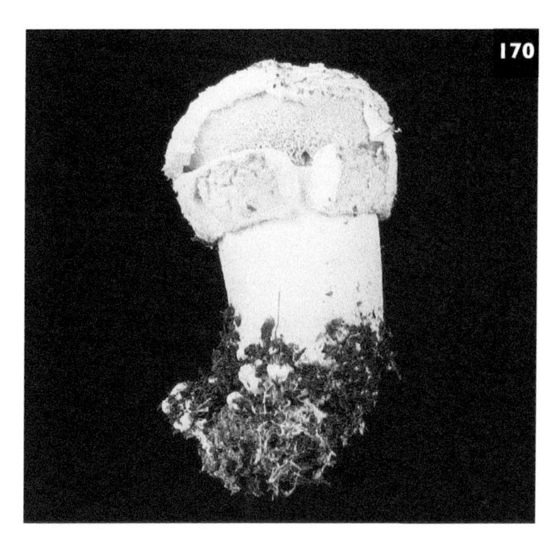

170 Hard gill may be related to a very dry atmosphere at the time of opening. The gills almost totally fail to develop.

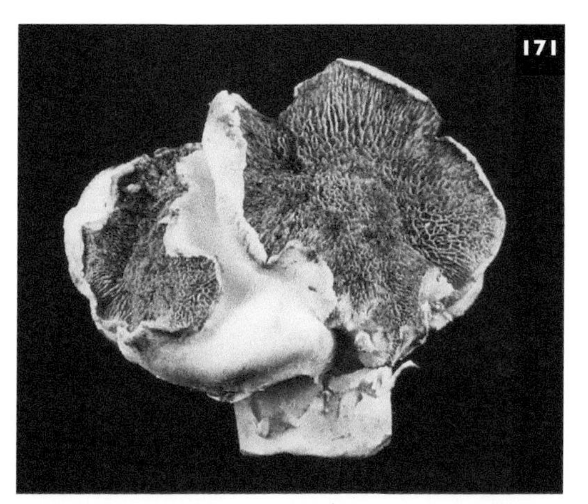

171 Rosecomb – gross distortion of a mushroom occurs when the gill tissue is formed on the top of the cap instead of underneath. This symptom is caused by chemically contaminated casing.

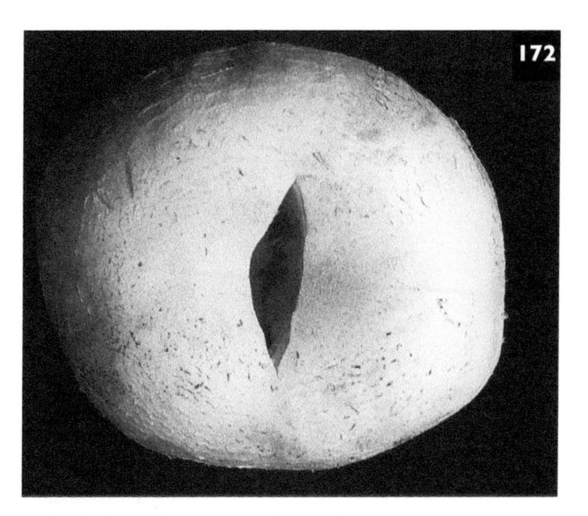

172 Split in the cap surface of a mature mushroom, probably due to an uneven growth rate.

Waterlogging

Small water-soaked areas in the developing mushrooms, especially in the first flush, are the commonest symptom. Clear patches appear on the surface of the mushrooms a day or so before harvest. Water-soaked tissue may also occur in the stems as transparent lines (**173**). The continuous dripping exudation of water from the edges of mushrooms (such mushrooms are referred to as 'Weepers') is not uncommon. Affected mushrooms may be in small groups within the crop, or can be individual mushrooms apparently randomly distributed within a normal crop. Droplets of water may occur on the surface of the stalks and the caps (**174**). Waterlogged stalks and caps are commonly found in vigorous large crops. Following harvest the waterlogged stalks often discolour, turning brown and eventually black. Water can be freely squeezed from such affected mushrooms.

General waterlogging is the result of an imbalance between excessive water uptake over water loss. Many factors may be involved, including the watering regime, the wetness of the casing, the temperature of the compost, and all aerial conditions that influence water loss. Many growers have to accept some degree of waterlogging as part of the penalty paid for higher yields. Recent research suggests that 'Weepers' in brown mushrooms may be governed by genetic factors similar to those described for Clustering.

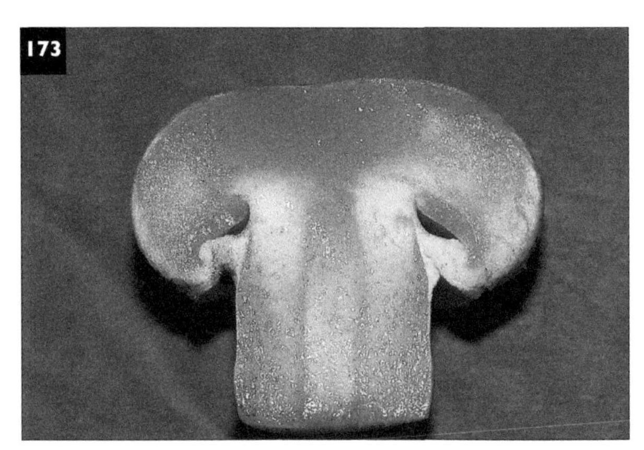

173 Waterlogging of a mushroom ready for harvest. Both cap and stalk are saturated.

174 Weepers (**a**) and (**b**). Water droplets oozing from mushrooms ready to be harvested. Sometimes the droplets continue to form and drip onto the casing. Such mushrooms are called weepers. The droplets vary in colour from amber (**a**) to red (**b**).

Mushrooms at harvest and post-harvest

Dirty mushrooms

When mushrooms initiate too deeply in the casing, they emerge with casing on them (**175**). Deep initiation is due to the mistiming of initiation, which can be the result of incorrect cultural conditions, or associated with the wrong use of casing inoculum or cacing. If spawn-runs are uneven (*see* Overlay p. 167), some parts of the crop are ready to pin before others, and this can result in deep pinning in the backward areas. Drying the surface layer of the casing is also a possible contributor to deep pinning. The mycelium does not then grow into the upper drier layers of the casing.

Dirty mushrooms are avoided by changes in crop management.

Scaling or feathering

This is the natural reaction of the mushroom cap to excessive evaporation, and usually results from the movement of air over the cap surface (*see* **46**). Dry air at very low air speeds does not cause the disorder, nor does high-speed air movement with wet air; the problem arises when the speed is out of balance with the water content (*see* p. 58 and **95**).

If scaling is a persistent problem the air distribution and humidification systems must be examined.

Browning

Staining of mushroom caps, when not caused by a pathogen (a bacterial cause is the most likely, *see* p. 108), can result from chemical damage. Leakage of phenolic vapour, or Formalin, from areas disinfected, or the incorrect use of pesticides, are possible causes. Generally the distribution of the affected mushrooms and sometimes the type of symptom differentiates this type of browning from that caused by pathogens.

175 Dirty mushrooms formed too deeply in the casing so that some small fragments of casing remain on their surfaces at harvest.

Bruising

Brown discolouration often develops following mechanical damage of the mushroom tissue during or after harvesting (**176**). Research work has demonstrated that water, both in the casing and in the compost, the humidity, the casing composition, the flush, and the mushroom strain, all play a part in the vulnerability of the mushrooms to bruising. Factors found to have the greatest effect include the casing wetness, the water content of the air and therefore evaporation, and the flush number. For instance, first flush mushrooms grown in wet casing did not bruise readily. Third-flush mushrooms showed least bruising when grown in drier casing. Watering treatments did not affect the incidence of bruising in second-flush mushrooms. Calcium chloride irrigation reduced the risk of bruising, but also advanced senescence. Some strains were found to bruise more easily than others.

It is likely that bruising is related to the anatomy of the mushroom. For instance, large cells with thin cell walls are more likely to be damaged and a higher calcium content of the cell wall may decrease damage. Calcium content of the cells is related to water uptake, and the thickness of the cell walls is related to the amount of cell expansion. The grower may not be able to predict the way the environment on a farm will affect these basic factors. Sensitivity to bruising appears to be part of the penalty paid for large yields.

176 Mushrooms at the edge of trays which have become damaged as a result of harvesters almost standing on them when higher trays are harvested. Standing on the edges of trays is not good practice, as it can result in the spread of pathogens.

Further Reading

Chang S.T., Hayes W.A. (1978). *The Biology and Cultivation of Edible Mushrooms*. Academic Press, London.

Chang S.T., Miles P.G. (2004). *Mushrooms: Cultivation, Nutritional Value, Medicinal Effect and Environmental Impact*. CRC Press, Boca Raton, Fl.

Flegg P.B., Spenser D.M., Wood D.A. (1985). *The Biology and Technology of the Cultivated Mushroom*. Wiley, Chichester.

Griensven L.J.L.D. van (1988). *The Cultivation of Mushrooms*. Darlington Mushroom Laboratories Limited, Rustington, Sussex.

Huang Nian Lai (1993). *Edible Fungi Cyclopedia*, published in Chinese. ISBN 7-109-02614-0/S.1681.
Hussey N.W., Read W.H., Hesling J.J. (1969). *The Pests of Protected Cultivation*. Arnold, London.

MacCanna C. (1984). *Commercial Mushroom Production*. An Foras Taluntais, Dublin.

Rinker D.L. (1993). *Commercial Mushroom Production*. Ontario Ministry of Agriculture, Food and Rural Affairs Publication, 360.

Vedder P.J.C. (1978). *Modern Mushroom Growing*. Educaboek, Culemberg.

Wuest P.J. (1982). *Pennsylvania State Handbook for Commercial Mushroom Growers*. The Pennsylvania State University, Pennsylvania.

Glossary

Abiotic disorder A disorder caused by a physical, chemical or environmental factor (*cf* Biotic disorders).
Absolute filter An air filter capable of removing particles down to 2 microns.
Absolute humidity The quantity of water held in a kg of air at a specified temperature (*cf* Relative humidity).
Aerobic With oxygen.
Agar gel column Agar gel columns are used in analytical processes where it is necessary to separate chemicals with different sized molecules. The process is used in the PAGE test for virus identification.
Aleuriospores Cells formed from fungal mycelium which develop thick walls and are able to withstand adverse conditions; aleuriospores are formed terminally.
Ambient temperature The natural (outside) temperature at the time.
Anaerobic Without oxygen (*cf* Aerobic).
Anamorph The imperfect (asexual) sporing state of a fungus.
Anastomosis The growing together of fungal mycelium and the fusion of the cell walls in such a way that cell contents can move from one individual to the other.
Antenna A (usually) much jointed, whip-like, mobile, sensory appendage to the head of an insect.
Antiserum Proteinaceous materials (*antibodies*) with properties of chemically recognizing other specific proteins (*antigens*).
Apothecia One of the sexual spore-producing structures of a group of fungi called the Ascomycotina.
Ascocarps Structures formed in apothecia and other structures in which ascospores are produced.
Ascospores Spores produced in ascocarps and other structures as a result of sexual reproduction in a group of fungi called the Ascomycotina.
Bacilliform Cylinder-shaped.
Bacteriophage A virus infecting a bacterium.
Balloon syndrome Describes an unusual symptom, *see* Chapter 9.
Biocontrol, biological control Usually refers to the control of pests or diseases using a biological means rather than chemical.
Biomass The biological debris produced during composting and deposited on the compost matrix. It consists predominately of dead bacteria.
Break see Flush.
Biotic disorders Are caused by organisms and viruses.
Bubble A distorted mushroom in which normal differentiation has not occurred and the contorted mass of tissue resembles a series of spheres or bubbles.
Button A button mushroom (2–3 cm diameter) has reached the developmental stage where the cap and stalk are fully differentiated and partly expanded but the tissue covering the gills is not broken, so that they remain enclosed.
Cacing Spawn-run compost added to casing at the time of applying the casing.
Casing A mixture of peat (or other nutritionally inert organic material) and an alkaline material such as chalk, limestone or sugar beet lime which is placed on top of compost usually when fully colonized by mushroom mycelium.
Casing inoculum or CI Mushroom spawn grown on a fragmented, low-nutrient medium, used to inoculate the casing at the time it is applied to the compost.
Case-run Colonization of the casing by mushroom mycelium.
Chlamydospores Cells formed from fungal mycelium which develop thick walls and are able to withstand adverse conditions (*cf* aleuriospores, which are terminally formed chlamydospores).
Conditioning A stage of compost preparation, after maximum temperatures have been reached in phase II, during which thermo-tolerant fungi colonize the compost and the ammonia level declines.
Cook-out or post-crop pasteurization Heat-treatment of the spent crop at termination which kills all mesophylic organisms within and on the crop and on surfaces of the structure in which it is being grown.
Conidia Asexual spores produced by many fungi, often on structures called *conidiophores.*
Differentiation A process that occurs during the development of a mushroom in which stalk and cap are formed.
Disinfect The process and outcome of the use of a disinfectant.
Disinfectant A chemical used to clean surfaces and rid them of unwanted organisms. In mushroom culture disinfectants are used to treat any surfaces on the farm that might harbour pathogen or mushroom spores and or pests.

Disease team or diseasers Workers on a mushroom farm with the specific function of treating and/or removing diseased mushrooms and also treating the contaminated area. This job is often referred to as *diseasing*.

DNA technology Methods used in the identification of nucleic acid. *DNA* (deoxyribonucleic acid) is the genetic material of the mushroom.

dsRNA the genetic component of mushroom viruses, which exists as two strands (double stranded or *ds*) of ribonucleic acid, and the site of the genes which characterize the virus.

ED_{50} and ED_{90} Used in the expression of the sensitivity of a fungus to a toxic substance such as a fungicide. Values are determined by growing the fungus on agar and also on agar amended with the toxic material, using a range of concentrations. The concentration that reduces growth of the fungus by 50% is known as the ED_{50} and by 90% the ED_{90}.

EM Electron microscopy. Has been used in the identification of mushroom viruses.

Endornavirus The proposed name for a group of viruses found in green plants and in some fungi. They are characterized by having double-stranded RNA genomes which vary in length from 14 to 17.6 kilobases; they are without a protein coat and they replicate within cytoplasmic vesicles. They occur in every tissue at every developmental stage and at a constant concentration of 20–100 copies per cell. In green plants they are transmitted in pollen and seeds and, with one exception, produce no known symptoms.

Enzyme A proteinaceous material capable of triggering or catalyzing a biochemical process.

Fastidious bacteria A group of bacteria extremely difficult or impossible to grow on an artificial medium.

Flush also known as break The appearance of a large number of mushrooms. Many commercial crops consist of two or three flushes.

Gills The tissue in the mushroom sporophore which produces the spores (basidiospores) of the mushroom.

HEPA High efficiency particulate air. A HEPA filter must remove at least 99.97% of all airborne particles by particle count, at a size of 0.3 microns.

Hygiene Cleaning and disinfecting to remove unwanted organisms.

Hyphae The individual filaments of the fungal mycelium.

Initials Small aggregates of mushroom mycelium which are the first stages of sporophore formation.

Inoculum A preparation of an organism which is used to *inoculate* another organism or medium.

Instar Stages in the development of fly larvae as they grow larger.

Integrated control The use of all means of control of unwanted organisms (*see* IPM).

IPM Integrated pest management, usually used in relation to pest control, in which all available means of control are integrated, i.e. chemical, biological, cultural, genetic, and any other.

ISEM Immunosorbent electron microscopy. This is similar in principle to EM except that the grids are first coated with a specific antiserum which catches, and thereby concentrates, the virus particles. This technique greatly increases the chances of finding virus particles in a mushroom preparation, providing a specific antiserum is available.

Knock-down Term applied to a rapid kill of flying insects.

Kill The period during phase II composting when the temperature of the compost reaches its maximum, which is usually 57–60°C.

Larva The pre-adult or immature stage of a pest.

Latent period The time between infection and symptom production.

Lesion A defined area of decay.

Levelling A relatively short period at the initial stage of phase II composting when the compost temperature is allowed to stabilize and become more or less uniform throughout.

LIV La France isometric virus is a spherical 36 nm particle with a number of dsRNA components, which show as nine bands when examined by the electrophoretic gel tests.

Macroscopic Seen by the naked eye without any artificial aid to vision.

MBV Mushroom bacilliform virus which is a cylinder-shaped particle more or less rounded at one end, measuring 50 × 19 nm. It is believed to play no part in the development of La France disease.

Mesophylic or mesophyl Organisms capable of normal growth in the temperature range of 10–35°C.

Micron (μm) A millionth of a metre.

Microscopic Visible only with the use of a lens or microscope.

Mould In this book the term 'mould' is used to describe fungi, other than the cultivated mushroom and its fungal pathogens, which occur in compost or casing at any stage of cropping.

Mucilage A sticky material, usually of a carbohydrate nature, which frequently surrounds spores of fungi and cells of bacteria.

MVXD Mushroom virus X disease sometimes also referred to as MVX. Unlike *LIV* or *MBV*, MVX does not refer to a virus particle but to a disease syndrome, and in this book the name MVXD is preferred for this reason.

Mycelium The vegetative body of a fungus, made up of hyphae.

Mycoparasitic The name given to an organism, (fungus, bacterium or virus) which is a pathogen of a fungus.

Mycophagous Feeding on mycelium.

Mycoplasma Microscopic structures of variable shape capable of inducing diseases in plants and animals, but so far not found in mushrooms.

Mycotoxic Toxic to fungi.

Nanometre (nm) One millionth of a millimetre.

Oospores The sexually produced resting spores of some fungi, e.g. *Pythium* spp.

Over-composting The condition of the compost where decomposition has proceeded beyond the optimum point for mushroom growth and has resulted in excessive breakdown of the organic material.

Oviposition Egg-laying.

PAGE A means of identifying nucleic acid based upon the separation of molecules of different molecular weight in a gel (agarose or polyacryamide) column. dsRNA fragments travel various distances according to their size and can then be identified using a stain. With La France disease nine bands are found and with MVXD up to 26. Generally, three bands of unknown significance occur in all hybrid strains of mushroom (*see Vesicle virus* p. 123).

Paedogenesis Asexual reproduction in the larval or pupal state.

Pasteurize A treatment which eliminates some organisms but not all. Used in the context of mushroom compost it is a temperature treatment which removes pests and pathogens, but leaves organisms that are required for the production of satisfactory compost.

PCR Polymerase chain reaction is a process for identifying and increasing a specific sequence of nucleic acid (DNA or RNA). The technique is particularly useful when very small amounts of nucleic acid are present and may, one day, become the standard tool for the identification of mushroom viruses and related diseases.

Peak-heat The stage in phase II composting when the maximum temperature is reached, also known as the kill.

Phase I an early stage in the preparation of mushroom compost after the initial mixing, when the ingredients are stacked in rows or put into bunkers.

Phase II The compost at the end of phase I is put into a purposely constructed tunnel or into growing containers in a room and further composted, initially at a temperature of 60°C, to eliminate unwanted organisms followed by the completion of the composting process.

Phase III This stage starts at spawning and continues until the compost is completely colonized by mushroom mycelium. The term is normally applied to spawn-running in bulk.

Phase IV This process combines case-running and initiation in specialized facilities before moving to cropping houses.

Phoresy/phoretic The transport of one organism by another. Usually related to the transport of mites by flies in mushroom cropping.

Pileus The cap of the mushroom.

Pin or pin-head A stage in the development of the mushroom at, or near the end of, differentiation of the cap, but before any enlargement has taken place.

Polyphage An organism that feeds on a wide range of other organisms.

Post-crop pasteurization see Cook-out.

ppm Parts per million, as a measure of concentration.

Prewet The thorough wetting of straw or other bulky organic materials immediately before phase I composting. This process is sometimes referred to as phase 0.

Pupa Quiescent, non-feeding stage during which a larva changes into an adult insect.

Relative humidity The degree of saturation of air. RH is expressed as a percentage at a particular temperature (*cf* absolute humidity).

Resistant Used to describe a pest or pathogen which has become able to withstand a concentration of pesticide which would be lethal to an unchanged or sensitive population; also used to designate a host which is only partially affected or unaffected by a particular pest or disease, commonly damaging to susceptible strains.

Retrotransposons Transcription of an RNA sequence into DNA which can then be inserted into a genome. Retroviruses in humans and animals also behave in this way.

RFLP Restriction fragment length polymorphism. Fragments of DNA which are used as markers to enable the identification of like sequences.

Rickettsia A type of bacterium. Those associated with diseases in plants are rickettsia-like and not strictly rickettsias.

RT-PCR Reverse transcriptase polymerase chain reaction is used to copy small fragments of virus RNA to give a detectable DNA product.

Ruffling The practice of the disturbance and distribution of mushroom mycelium in the casing after it has grown about a third of the way up the casing.

Salting The use of salt (sodium chloride) to cover diseased mushrooms or affected areas of casing.

Sanitize see Disinfect.

Sanitizer see Disinfectant.

Saprophage An organism that feeds on non-living organic matter.

Sclerodermoid mass An undifferentiated mass of sporophore tissue induced by the presence of a pathogen, in particular *Mycogone* (*see* Bubble).

Spray-off The application of a disinfectant at the end of a crop to kill pathogens on mushrooms and to reduce the contamination of the casing surfaces.

Spawn A packaged pure culture of a specific strain of *Agaricus bisporus* growing on a medium, usually grain.

Spawned casing Spawn-run compost, or specially produced spawn, mixed into the casing before or at application of the casing (*see* Cacing).

Spawning The introduction of spawn into the compost.

Spawn-running Colonization of the compost or casing by mushroom mycelium.

Sporangiophores The stalks of sporangia which bear sporangiospores.

Sporangia An asexual sporing structure of fungi in the Mastigomycotina.

Spore A general term used to describe various structures produced by fungi which are capable of germinating and reproducing the fungus.

Sporophore The spore-producing structure of some fungi, e.g. the mushroom.

Sticky trap A card painted with polybutenes, or similar sticky material, on which flies are trapped; used as a method of monitoring pest incidence.

Stipe The stalk of the mushroom.

Strain A distinct variety or type.

Supplementation The addition of a nutritional supplement at spawning or sometimes just before casing.

Sweat-out Archaic name for peak-heat or kill applied to phase II composting.

Teleomorph The perfect (sexual) sporing state of a fungus.

Thermal death-point The expression of temperature and time of exposure which will kill an organism (i.e. 60°C for 2 hours).

Thermo-tolerant Organisms that are able to withstand long periods at high temperatures. In mushroom culture, thermo-tolerant organisms survive the temperatures reached in phase II composting.

Transposon A genetic element having at least the genes necessary for its own transposition. Transposition involves the movement from a site in a genome to another site in the same or different genome. In simple transposition, the DNA is moved from a site where it leaves a lethal gap. Transposons may have a limited amount of genetic material, such as antibiotic resistance.

Under-composting The condition of the compost where the process of composting has not reached completion, and therefore the compost is not entirely suitable for mushroom growth.

Vector An organism that is essential for the transmission of pathogens.

Verticillate conidiophores A whorled arrangement of spore-bearing stalks which bear conidiospores.

Vesicle virus A virus with no known effect found in modern spawn strains in which naked RNA is contained within a vesicle. There is no associated protein coat.

Virulence The level of aggression of a pathogen towards its host.

Viroid Very small amount of ribonucleic acid without a protein coat, capable of causing diseases in plants but not so far found in mushrooms.

Virus Small disease-causing particles of various shapes with a nucleic acid component surrounded by a protein coat. In mushrooms the nucleic acid is generally double-stranded RNA, surrounded by a protein coat (*see LIV, La France isometric virus*).

Watering-tree Equipment used for watering a mushroom crop, consisting of a vertically mounted boom with short lateral branches fitted with nozzles.

White-line test A test for the identification of the bacterium which causes Bacterial blotch disease (*Pseudomonas tolaasii*). A related bacterium is grown on the same agar plate as the isolate to be tested and a positive result is marked by the occurrence of a white line of precipitation between the two organisms.

APPENDIX 1

Registered Pesticides

The number of pesticides registered for use on mushrooms is constantly under review and many products are no longer on National Lists. The tables in Appendix 1 contain the best available information, but are likely to be out of date in a very short period.

It is therefore vital to check the National List and read the product label carefully before using any pesticide on mushroom crops.

Pesticides registered for use on mushrooms in Australia and South Africa

	Country	
	Australia	South Africa
Fungicides		
Prochloraz manganese	*	*
Prochloraz zinc		*
Thiabendazole	*	
Bactericides		
Calcium hypochlorite	*	
Chlorine dioxide	*	
Insecticides		
Cyromazine		*
Diazinon	*	*
Dichlorvos	*	*
Diflubenzuron	*	*
Fipronil	*	
Mercaptothion		*
Mercaptothion/pyrethrin		*
Permethrin	*	
Triflumuron	*	
Nematocide		
Fenamiphos	*	
Wood treatment		
Propiconazole	*	

Pesticides registered for use on mushrooms in Europe

	Country								
	Germany	Belgium	Spain	France	UK	Ireland	Italy	Netherlands	Poland
Fungicides									
Benomyl			*	*				*	
Carbendazim			*		*	*		*	
Chlorothalonil				*		*			*
Iprodione			*						
Copper oxychloride			*						
Prochloraz manganese		*	*	*	*	*	*	*	*
Sodium hypochlorite					*				
Thiophanate methyl								*	
Thiabendazole						*	*		
Zineb						*			
Insecticides									
Bendiocarb			*						
Cypermethrin									*
Chlorfenvinphos				*		*	*		
Cyromazine			*	*			*		*
Deltamethrin			*			*	*	*	*
Diazinon		*	*			*			
Diclorvos			*						*
Diflubenzuron	*	*	*	*	*	*	*	*	*
Endosulfan		*							
Fenthion			*						
Lindane						*			
Malathion								*	
Methoprene				*	*	*	*		
Naled			*						
Nicotine						*			
Permethrin		*			*	*	*		
Piperonyl butoxide		*							
Pyrethrin		*			*				
Resmethrin					*	*			
Sulfotep		*	*	*		*			*
Teflubenzuron		*					*		*
Trichlorfon		*							
Nematode									
Steinernema feltiae					*	*		*	
Wood treatment									
Azoconazole		*			*				*

These data are based on an EU list dated May 2005.

Pesticides registered for use in Canada and the USA

	Canada	USA
Fungicide		
Chlorothalonil	*	*
Thiophanate methyl	*	*
Bactericides		
Calcium hypochlorite	*	*
Calcium chloride	*	*
Insecticides		
Bacillus thuringiensis var. *israelensis*	*	*
Azadirachtin		*
Cyromazine	*	*
Diazinon	*	*
Dichlorvos	*	
Diflubenzuron		*
Dimethoate	*	
Malathion	*	*
Methoxychlor	*	
Methoprene	*	*
Permethrin	*	*
Pyrethrin	*	*
Nematode		
Steinernema feltiae	No registration required	
Wood treatment		
Copper quinolinate	*	*
Propiconazole	*	*

APPENDIX 2

Checking the Chlorine Content of a Hypochlorite Disinfectant

Because the shelf-life of hypochlorite disinfectants is short, especially if they have not been stored in a cold dark place, it is important to check the available chlorine content at intervals, especially if large containers are bought and not used quickly.

Procedure

1. Prepare solution A by dissolving 12.5 g of potassium iodide in a 12.5% v/v solution of glacial acetic acid in water.
2. Prepare solution B by making a 0.28 M solution of sodium thiosulphate (this is done by dissolving 44.24 g in 1 litre of de-ionized water).
3. Measure 80 ml of solution A into a container, such as a conical flask.
4. Add 1 ml of the test hypochorite solution.
5. Add solution B by 1 ml lots, while shaking the flask, until the brown colour disappears.

For every 1 ml of solution B used to remove the brown colour, the test solution contains 1% of available chlorine. If the chlorine concentration is less than 10% this must be taken into account when using the hypochlorite disinfectant (*see Table 5*).

APPENDIX 3
Some Useful Conversions

Length
1 nanometre (nm) = 1 millionth of a millimetre
1 micron (µm) = 1 millionth of a metre
1 millimetre (mm) = 0.04 inches
1 centimetre (cm) = 0.40 inches
1 metre (m) = 39.40 inches or 3.28 feet or 1.09 yards
1 kilometre (km) = 0.62 miles

1 inch = 2.54 cm
1 foot = 0.30 m
1 yard = 0.91 m
1 mile = 1.61 km

Area
1 square centimetre (cm^2) = 0.16 square inches
1 square metre (m^2) = 10.77 square feet or 1.2 square yards
1 kilometre (km^2) = 0.39 square miles

1 square foot = 0.09 m^2
1 square yard = 0.84 m^2
1 acre = 0.40 hectares (ha)

Volume (dry)
1 cubic centimetre (cm^3) = 0.061 cubic inches
1 cubic metre (m^3) = 1.31 cubic yards or 35.31 cubic feet

Volume (liquid)
1 fluid ounce = 28.41 ml
1 pint (Imp) = 0.57 litres
1 gallon (Imp) = 4.55 litres
1 gallon (US) = 3.79 litres
1 millilitre (ml) = 0.035 fluid ounces
1 litre (l) = 1.76 pints or 0.22 Imp gallon or 0.26 US gallon
1 US spawn unit = 1 litre

Weight
1 ounce = 28.35 g
1 pound = 453.6 g

1 gram (g) = 0.035 ounce
1 kilogram (kg) = 2.21 pounds
1 ton = 0.91 tonne

Speed
1 metre per second = 3.28 feet per second

Temperature
$°F = (°C \times 9/5)$
$°C = (°F - 32) \times 5/9$

Index

For Product Safety Concerns and Information please contact our EU representative GPSR@taylorandfrancis.com Taylor & Francis Verlag GmbH, Kaufingerstraße 24, 80331 München, Germany

T - #0319 - 160425 - C192 - 260/192/10 [12] - CB - 9781840760835 - Gloss Lamination